Handbook of Minor Fruits, Medicinal and Aromatic Plants

Handbook of Minor Fruits, Medicinal and Aromatic Plants

Edited by:

Esha Rami

Delve
PUBLISHING

www.delvepublishing.com

Handbook of Minor Fruits, Medicinal and Aromatic Plants

Esha Rami

Delve Publishing
224 Shoreacres Road
Burlington, ON L7L 2H2
Canada
www.delvepublishing.com
Email: orders@arclereducation.com

ISBN: 978-1-77469-080-2 (Hardcover)

ABOUT THE EDITOR

Dr. Esha Rami, is presently working as Assistant Professor & Head, Department of Biotechnology, Parul Institute of Applied Science, Parul University, India. She did her Post graduated and PhD in Biotechnology from Ganpat University. She has authored a number of national and international publications.

TABLE OF CONTENTS

LIST OF FIGURES

Figure 1.10: The fruit has a similar look when pelted with lychees and long fruits. The sweet but creamy taste and seeds in its midst are its translucent white shell. Rambutan is highly nutritious and can provide medical benefits that range from initial weight loss and improved metabolism to increasing disease resistance.

Figure 1.11: While not juicy, the fig is a crop with an unbelievably luscious fragrance and a pleasant taste. Initially from Asia, figs now thrive in the Mediterranean, and thousands of hybrids are divided into four principal colors: black/purple, red, and white.

Figure 2.1: Pollution by plants is the major cause of environmental degradation which will be most harmful to future generations.

Figure 2.2: Majority of the minor fruits are produced during the summer seasons in the months of May, June, July and August.

Figure 2.3: Bananas are minor fruits that are quite easy to cultivate and require minimal input and effort. they can be grown at the home garden which makes them very useful, especially to resource poor individuals.

Figure 2.4: The Garcinia species grow extensively in a semi-wild state in India and are considered important in the northeast regions.

Figure 2.5: Resource poor individuals in the rural areas need to be educated on the production of minor fruits in order to be able to improve health wise and economically.

Figure 2.6: Avocado fruit is a good source of fat that is not harmful to the human body but can be very useful.

Figure 2.7: Bael fruit, commonly known as wood apple fruit has the highest carbohydrate content compared to the majority of minor fruits.

Figure 2.8: Most fruits are useful in the prevention and dealing with diseases. berries for instance are useful when dealing with arthritis.

Figure 2.9: Jack (*Artocarpus heterophyllus*) plant and fruits.

Figure 2.10: Custard Apple (*Annona* spp) used for medicinal purposes of curing patients

Figure 2.11: Jamun Plum (*Syzygium cumini*) source of iron, sugar, minerals and carbohydrates.

Figure 2.12: Ber (*Ziziphus mauritiana*) be used in the preparations of jam and wine

Figure 2.13: Passion fruit (*Passiflora edulis* Sims) is used medicinally for urinary infections and as well as a mild diuretic, digestive stimulant, and heart tonic and to treat asthma, whooping cough, bronchitis, coughs, gastric cancer ,etc.

Figure 2.14: Drum Stick (*Moringa oleifera*) to have 7 times the vitamins C in oranges, 4 times the calcium in milk, 4 times the vitamin A in carrots, double the proteins in milk and thrice the potassium in bananas.

Figure 2.15: Pumelo fruit rich in potassium, which helps regulate fluid balance and blood pressure.

Figure 2.16: Macadamia (*Macadamia integrifolia*) contains benefits which include improving digestion, heart health, weight management and blood sugar control.

PREFACE

Various minor fruits are available today for consumption. This volume addresses the achievements, limitations and prospects that minor fruit products provide in various countries. A number of tribal groups depend on minor fruits for sustenance. Minor fruits also fundamentally promote the development of herbal medicines. Minor fruit trees, such as velvet apples, jujube and cowa provide a means to prevent hunger among rural communities. Minor fruits can be of great help in maintaining the promise of hunger alleviation and providing nutritional value.

The volume points out that in coastal areas, the production of the minor fruit ber (jujube) has made a huge contribution in reducing poverty and restoring soil from saline to non-saline state. There are always some areas where food shortages, nutrition concerns and poverty are high. In addition, there are occasional food emergencies due to lack of employment, and poor crop harvest. When traditional fruits such as avocado and oranges are scarce, due to calamities like drought, plant disease or flooding, minor fruits can make a great alternative.

Minor fruits grow wildly in unattended places such as roadsides and wastelands, but they are safe and healthy to consume. These products are less popular than their mainstream counterparts, plus with less market interest. They are not grown commercially like apples or oranges, but they are very tough, often developing well even in fragile soils and environments where other fruits can't grow.

This volume covers a number of topics including; chapter 1: minor fruits, chapter 2: role of minor fruits in food and livelihood security, chapter 3: value addition in minor fruit crops, chapter 4: processing of minor fruits, chapter 5: medicinal and aromatic plants, chapter 6: history of medicinal and aromatic plants, chapter 7: cultivation of medicinal and aromatic plants, chapter 8: regulation of medicinal and aromatic plants, and chapter 9: quality, advertising, and labeling of medicinal and aromatic plants.

Minor fruits are less affected by pests and diseases compared to major fruits. They grow spontaneously since they are not cultivated by anyone and rely on wild cross-pollination to propagate. These have been naturally selected to adapt to any environment, are nutritious, and occupy an important position in home-grown medicines. Although there are a large number of natural fruit products in different agricultural biological areas, there are actually no coordinated plantations and no detectable cultivars. Minor fruits are highly underutilized, and some of them are rotting away in the wild because there is no one to harvest them.

Among these fruits, ascorbic acid is present in the highest concentration in wood apples but the lowest concentration in roselle. The most notable elements in Monkey Jack fruit were carotenoids, zinc and copper. Calcium, magnesium and phosphorus were high in *Antidesma velutinum*. Potassium has the greatest content in Wood apples, closely followed by Monkey Jack. It is worth noting that most of the minor natural fruits have higher amounts of ascorbic acid compared to other fruits. Most minor fruits have higher nutritional and mineral content than major fruits. People are urged to grow more minor fruits to protect these fruit trees from going extinct.

Chapter 1 MINOR FRUITS

CONTENTS

1.1 INTRODUCTION

Minor fruits are currently a category of fruit that grows on sidewalks, residences, wastelands, etc. scattered and with little to no attention. Generally speaking, the fruits are eaten by humans, but comparatively less palatable than other essential fruits that have lower consumer value are called minor fruits, but are not cultivated to a restricted degree in coordinated plantations with the use of artificial agri-inputs. They are generally robust and mature well even in weak soil and environment with intense exploitation capacity. In most of such crops, attacks on fatal insects and pathogens are comparatively fewer. They naturally reproduce and expand instead of becoming obsolete. Since there is a negligible area below each of such fruit trees, they are generally recognized as 'minor fruit.' In India, many underused fruits grow mainly in homesteads, jungles and forests and on roads and railways lacking much consideration. These are suited to the local environment, are rich in nutrients and lead to hunger and food security in rural areas in the home. Even though in different Indian agricultural production areas, there is a large variety of underused fruit crops, for nearly all these minor or under-used crops there are no coordinated orchards and no established crop yields and some are degraded by the plant-rich list (Altendorf, 2018).

Figure 1.1: Drumstick, ber, Jamun, custard apple and jackfruit, among the significant little tropical natural product developments, are superb sources of fundamental nutrients, minerals and anti- oxidants. Jackfruit is a yielding organic product harvest of rich dietary benefit (*Artocarpus heterophyllus* Lam.)

Source: *https://www.indiamart.com/proddetail/cultivation-of-minor-fruits-practices-and-principles-book-19307053812.html*

For every three people in the world, malnutrition impacts one child. It is projected that the overall number of children below five is 159 million. For one or multiple essential nutrients two billion individuals are impaired and the figure is about the same in obese or overweight. Variety in food plates enhances diets by including nutrients that are not available. Unused small fruits will fill the gap because they thrive on all sorts of soils in unfavorable situations. In a country, like India in which most of the growers are capital-poor and grow under less favorable circumstances this intensity of minor fruits is especially helpful. Below we will discuss some of the common minor fruits.

1.2 LITCHI

The Litchi's origins are from southern China, mainly the Fukien and the Kwangtung provinces. Possibly because of its soil, environment and the short lifetime of its seed, the expansion of the litchi to certain areas of the planet was very sluggish (Banyal et al., 2019). Within the 18th century, Litchi entered India via the North-East and Myanmar. India is after China as the second-largest litchi manufacturer in the world. Florida, Madagascar, South Africa, Australia and Thailand are several significant manufacturing nations.

Litchi is ranked 7th in area and 9th amongst fruit crops, however, in India, it is ranked 6th in cost. The litchi's national mean production is 6.1 t/ha, which is significantly less than a well cared for crop yield. Litchis have an overall production of 10.5 tons/ha for western Bengal and 8.0 tons/ha for Bihar. In others, the production in Uttaranchal is slightly lower than 1.0 t/ha.

Litchi's alimental value is primarily dependent on its variety-by-variety sugar content The fruit is also rich in Vitamin B1, Riboflavin, carotene, iron (0,03%), phosphorous (0,32%), calcium (0,21%), fibrous substances (2,25%), minerals (0,7%); carbohydrates (9,4%), fats (0,3%) and proteins (0.7%), and fibrous materials (0.1%). Litchi produces a really fine canned fruit.

The litchi fruit which is utilized in the summers, also produces a rather flavored squash. Several certain items are made with Chinese litchislike pickles, preserves and wine. The Chinese sometimes refer to dried litchis as the litchi-nut.

Sub-tropical fruit Litchi is ideally grown in humid sub-tropical climates. It typically favors low heights and can be cultivated up to 800 m above sea level. Deep, adequately drained and organically enriched loamy soils with a pH range from 5.0 to 7.0 are suitable for cultivation. Winter frost and summer dry heat are barriers to its effective growth. For many years, growing trees need defense against frost and warm winds, even though such fluctuations in temperature are appropriate for properly growing trees.

In the North in the foothills of the Himalayas, from Tripura to Madhya Pradesh plains, Uttar Pradesh, and Jammu & Kashmir in India, economic agriculture has historically been limited. Even so, with a rising market and crop viabilities in many countries like Chhattisgarh, Madhya Pradesh, Jharkhand, Bihar, etc., commercial crops have spread.

The ground is completely lifted and stripped from all plants. The soil is then packed with a harrow and a disc plow. The growth of the litchi plants is negatively influenced by the hot and saturated wind in winter cold and summer winds (Barua et al., 2019). Consequently, an appropriate wind breach must be installed at the orchard boundaries for the appropriate development of litchi cultivation and must be at a right angle towards the prevalent wind direction Appropriate to this are the high-growth trees such jamun, mango, arjun, eucalyptus, etc.

The most popular way of air layering is propagation. It is normally achieved at the start of the monsoon, and after 60-70 days, the air layers can be separated from the main crop. Upon harvesting, the litchi plant also needs to be made through a considerable extent of pruning. Normally, no pruning is needed once the target shape is reached, excluding removing dead or diseasing branches and broken shoots or twisted extremities. The fruit part of the shoot shall be extracted at the harvest period, to encourage further growth.

If vegetative growth is excessive then root and shooting must be performed. Strong pruning, however, results in abundant vegetative growth, at the cost of flora and fruit. If the trees get too mature and grow small-scale fruits, we recommend severe pruning. During colder months, two watering are needed for the use of litchi trees at a frequency of 45-60 days. Watering is particularly needed within the hot summer months in quarters after fruit production between March and the conclusion of June The drainage basin or ring network is advisable. In addition to being economic, the usage of drip irrigation has proven good for a good yield.

Figure 1.2: There are around 200 cultivars, with types that develop at an early and late date and that are adjusted to a hotter and colder environment, yet in China, eight are essentially utilized for exchange. They are additionally developed for their natural product, as a fancy tree.

Source: *https://en.wikipedia.org/wiki/Lychee*

1.3 MANGOSTEEN

Garcinia mangostana L. commonly referred to as the mangosteen tree is from Southeast Asia. In honor of the research of the French botanist Laurent Garcin in the eighteenth century on the plant, Linné provided it its generic name Garcinia. It is cultivated in southern India and the southern Philippines, but mostly in the southern part of Indochina and Indonesia. Alone, in Malaysia and Thailand, almost 100 000 tons are planted on almost 25 000 hectares (Bisht, 2020). Its popularity has also ensured the fruit is widely spread all through the tropics and it is now found in Australia, Madagascar, Brazil and Central America in Queensland (with production exported to the United States). In the French overseas, divisions of a few specimens are identified but mangosteens are not seriously grown in these areas.

The tree develops to 10-15 meters in height. It has a wonderful slender pyramid pattern, which gains volume over time. The top portion of the leaves is greenish-yellowish and matte on the bottom. The flowers emerge at the end of the last year and have four waxy, tough petals. The fruit is round with a thick smooth corky skin, slightly flattened on the top and on the bottom; this is pale green to yellow, with purple violets turning as the result of the ripe fruit. The fruit inside comprises of six to eight white parts, which, like the mandarin, are divided by one or more seeds.

Mangosteen is a deciduous plant needing heat and humidity. It is important that average rainfall over 1 500 mm, spread well over the whole year. But for floral initiation a limited duration of drought presents advantages. The ideal production temperature is between 25 and 30°C. The young trees rarely stand up to the intense sun and are thus planted with shady accompanying crops. In the earlier years, it can be a reasonable option to combine trees with bananas, as bananas include a required shade. However, it is important to prevent any competition on nutrients and water from the accompanying crop.

Seeds make multiplication simple even if they are never actual seeds because they are produced lacking pollen and thus without pollination! There is a type of plant propagation. Seeds taken from the fruits have a rather minimal germinating ability. They must be planted rapidly or germinated before a shooting occurs in damp cotton wool. The seeds can then be carefully transplanted into a pot with a light organic material. Stepping is feasible, but it is also very challenging since it needs to be done in a saturated setting (Behera et al., 2016). Depending on the process of propagation, this step in a shade house is lengthy as it requires two to three years to mature mangosteen that is ready for plantation.

While many variations across the world have been registered, there appears to be only one major mangosteen type. After 8 to 10 years the tree develops very gradually and the first fruit is produced. This long non-productive time discourages many farmers. For that reason, the plant is hardly ever cultivated with other trees, including rambutan and durian, in orchards of its origin area. A tree can grow between 200 and 2 000 fruits in adults. The typical yield in Asia nevertheless amounts to between 4 and 6 tonnes/ha. Two or three days following changing color, the fruits are chosen—from green to purple. The ultimate value of the fruit is determined by this harvesting process. The taste of the fruit is affected by selecting too fast (greenish-yellow peel with no rose spots).

Inappropriate circumstances like inadequate drainage or mineral shortage (primarily calcium and zinc) can influence the quality of the fruit and render it unacceptable for sale (translucent pump). Similarly, some pests cause severe negative impacts: fruit flies or bugs make yellow latex flows that render the white flesh bitter if touched. Thrips tend to be key mangoes and frequent pricking influences strongly the color of the fruit. The fruits will always remain at the average temperature for several days at the post-harvest level, while they will remain for three weeks at 10-12 °C (85-90 relatively high humidity) without affecting the value Savings caused significant physiological harm to temperatures below 10 °C d. Finally, while mangosteen pericarp tends to be good, there is a lot of loss in fruit processing.

Whenever the fruit is ideally mellow and newly sliced, the sweet treat of mangosteen flesh remains unmatched. The fruit is chosen in Thailand when it becomes pink (de 5 to 50 percent of the surface). They are carefully released so that latex does not enter the pulp and put into an acidic juice for 30 minutes to avoid brewing. Thais like this a lot, although the length of the item is restricted to five hours, which involves daily planning. It may also be combined with Mangosteen. Frozen whole fruit and juice, dried fruit, and fruit jelly are the primary items. The aforementioned are highly valued for their powerful antioxidants and supposed role in human wellbeing, for example, the reduction of serum fat levels (Boyd et al., 2007). The mangosteen juice's color becoming too pale, a pericarp colorant is applied. This provides a rose-tone and tannins that are healthy too. The fruit skins and the tree's bark also produce these tannins. It has anti-inflammatory features and is shown to be effective in treating amebic dysentery. Mangosteen wood is strong, hard, and red. It is valued in the development of cabinets, but seldom used because of its value!

1.4 STRAWBERRY

Because of its unrestricted inclusion and diversity in various diets, Strawberry's prominence proceeds to grow; *Fragaria vesca ananassa* is the most cultivated species. There is a small shoot, leaves, trusses and stolon in the plant that grows over the soil for a long time. In almost all areas of the world, strawberries will develop but water and warm temperatures must be constantly supplied. This plant can lso b grown in vessel if no land is accessible. Currently recommended varieties include Aiko, Doiglas, and Chandler.

Figure 1.3: The Fragaria strawberry is one of the world's most common fruits. More than 10 Fragaria species vary in texture, size and taste, but have the same standard regal, little green leafy stones with heads, seeded cover, red flesh and heart shapes.

Source: *https://www.bbcgoodfood.com/howto/guide/ingredient-focus-strawberries/amp*

The plant necessitates well-distributed precipitation of at minimum 1200 mm annually, but 25 mm a week or watering is required in areas that are not reachable. Slightly rooted, strawberry plants could thrive in a broad variety of soil types. Deep sandy flooring, nevertheless, is fine. The soil must be humus-rich, well-draining, and pH 5.5-6.5. The optimum altitude is from 1500 m to 2200 m above sea level while development in various areas can occur.

Organic matter is important for achieving high returns. Their soil water potential is increased, their framework is preserved and strengthened and so quick and thorough rooting is encouraged. DAP fertilizer and CAN be placed during the planting process. The soil should be filled to receive a loose and deep bed. Weeds should be destroyed and manure and ash applied. Get seedlings or scatter them. 3000 seedlings are needed for 1/8 acre land. Place strawberry seedlings amongst plants within 30cm and between rows over a distance of 45cm. Make sure that you water regularly (Chakraborty et al., 2011). Make sure that the field is properly sanitary. The application of any plastic mulch will accomplish this. Tap old and ill leaves daily.

The first fruit is produced and matured within approximately 70 days. They are chosen in full or almost red color, while partially colored berries color eventually but luckily will have complete flavor. Be careful not to hurt the berries. One in eight acres of strawberries produces 30-50 kg a week. Since the fruit can be highly consumed only after 4-5 days of harvest, one should search the market fast enough to prevent losses. Strawberries from the gardens can be consumed directly. In large metropolitan markets, they have higher competition, particularly from food providers and supermarkets. Added value involves: manufacturing of yogurt, manufacturing of ice cream, cake baking, producing jams among others.

1.5 PASSION FRUIT

The Passiflora sp.(passion fruit) belongs to the Passifloraceae family. It is a permanent plant that bears fruit in a year. In Meru, Murang'a, Kakamega, Nyeri, Kisii, Nyamira and Thika, passion fruits are grown amongst other areas in Kenya. It is possible to consume fruits freshly or turn them into juice. The two main varieties include:

- *Passiflora flavicarpa* – Once ripe, it's golden, in-shaped oval and not aromatic. The variation is suited to the tropics and therefore in Kenya extremely well. Before maturing, both types are normally green.
- *Passiflora edulis* – When mature, very acidic, and delicious and juicy. It is purple in color. It has a heavy herbal fragrance and is formed oval. Subtropical regions are ideal for this variety.

The plant of passion fruit prospers with a pH range of 6.0-7.0 and in deep well-dried soils. Even organic material should be contributing to the richness of the soil. The crop performs well between 1200-2000m ASL in altitude. Yearly precipitation of about 900-2000 mm is ideal for sustainable crop yield. Excessive precipitation contributes to inadequate nutrition and also causes sickness. A temperature range of 18-28 degrees centigrade is needed for production. Regarding propagation, cuttings, grafts, and seeds are propagated. First, in nursery beds, seedlings are raised. Seeds of consistency from the best fruit should be collected (Challies & Murray, 2011). Covering can be carried out to minimize the intense sunlight on the small seedlings and to shield the developing site from heavy wind. Passion fruits are climbers, and so it is necessary to create a trellis for support. Take-off, which requires frequent elimination of tendrils, allows the wines to intertwine to ensure that the side parts are freely suspended.

1.6 BILIMBI FRUIT

Bilimbis are incredibly bitter fruit with a yellow body and chewy juicy flesh. Bilimbis are very juicy. There are 5 distinct ribs on the oblong-shaped fruit. A five-speed star in a cross-section of the fruit highlights the pentagonal fruit shape. The fruit looks like gherkins with smooth skin. They grow on bushy trees beautiful red-purple flowers and green leaves in clusters. Every fruit may be made up of a number of flat, small, pale seeds. Bilimbi fruit is also known for its tart-tangy flesh as Belimbibg and Tree Cucumbers.

All year, bilimbi fruit is accessible. Bilimbi fruit is a native fruit, which is *Averhhoa bilimbi* in botany. The Bilimbi, a domesticated plant, is strongly linked to the starfruit. The fruit juice of Bilimbi has significant quantities of oxalate, a plant natural acid. Oxalic acid, or oxalate, is the typically acidic flavor of the Bilimbi fruit. Oxalate, however, can lead to an elevated likelihood of kidney stones and eventually renal failure if ingested in excess. The fruit trees of the Bilimbi are tiny and develop to around 15 meters. Each tree can grow hundreds of fruits, with an annual production of approximately 50 kilograms.

The vitamins A and C, and potassium are plenty in bilimbi fruits. In addition, studies have shown that the fruit has good antioxidant, antimicrobial and anti-inflammatory properties (Crush & Frayne, 2011). In treating and preventing live disease, cancer, stroke, diabetes and heart disease, Bilimbi fruit could theoretically be of use.

Bilimbi fruit is also processed into preservatives, relishes, condiments, and pickles. The fruit is also accustomed to add a bitter style to soups, curries, and sambal (a Southeast Asian chili paste). The fruit pairs well with certain meats like pork, shrimp, and fish. In rural elements of the Philippines and in geographic region and province in the Republic of India, the fruits are consumed raw – they normally can be enjoyed alone, or with sodium chloride. The tree is usually utilized in fashionable cooking, pairing well with flavors like mint, lemon rind, orange, star anise, and bark. The Bilimbi fruit may also be held on to for up to at least one week within the icebox.

Figure 1.4: In addition to its refreshing fragrance and flavor for the bilimbi fruit, it provides various medical advantages such as diabetes control, hypertension treatment, hemorrhoids correction, bone reinforcement and cough and cold relief.

Source: *https://specialtyproduce.com/produce/Bilimbi_Fruit_14716.php*

Bilimbi fruit is spread wide in ancient people drugs round the plant. In Asian nations, a concoction of the fruit and leaves were utilized to treat certain STIs such as syphilis, whereas a drink made of the cooked flowers, fruit and leaves was utilized to facilitate the curing of coughs. The juice of the fruit was utilized as a wash for the eye, and to assist in treating pimples (Chazovachii, 2012). The juice of the fruit is acidic and thus was additionally used as a house improvement substance; within the nineteenth century within the Philippines, it had been used as hand soap. The Malay individuals additionally once discovered the juice to be efficient in the removal of rust from their ancient keris dagger blades.

In the tropical areas all over the planet, the Bilimbi trees are identified. It is believed to have come from Malaysia or Indonesia. Bilimbi trees in house landscapes in South East Asia are frequently found. It is often grown in several areas of South and Central America and the Caribbean, where it is called Mimbro. In Australia, it is also cultivated commercially. Bilimbi fruit has historically been utilized for the management of hypertension and diabetes for medical purposes. In the 14th century, Egypt recorded the earliest history of the plant.

1.7 CARAMBOLA

Averrhoa carambola (carambola tree) is a woody crop in the Oxalidaceae family cultivated for its fruit called starfruit. Carambola trees with falling branches are bushy, small evergreens. The tree has smooth, composite leaves consisting of 2-5 green colored leaflets, arranged around the branches in a spiral manner. Every leaf is 15–20 centimeters in length and every leaflet is 3.8–9 centimeters in length. The flyer is ovate or oval shape.. The fruit is waxy and orange-yellow, length 7.5–12.5 cm, oval and profoundly ribbed, produced by tree tiny lilac or violet flowers. The tree has a 40 years long lifespan and has a diameter of between 6–7.6 m (20-25 feet), extending to between 7–10 m (22-33 feet) in height. Carambola can also be referred to by Southeast Asian names as coromandel gooseberries, five corners, starfruit, and Averrhoa. The fruit can be consumed cooked or fresh, and it is possible to use the juice in an iced drink. It can also be utilized as a seasoning in relishes (Chakraborty et al., 2011).

Carambola is subtropical to a tropical tree that is best grown between 20 and 35°C (68–95°F), in hot to warm temperatures. Trees can usually no longer grow under 18.3°C, however, existing trees will stand at -2.8°C (27°F) for brief times. Carambola, from sandy soils to heavy clay soils, can be cultivated effectively in a variety of soil varieties but is appropriately developed in a healthy, well-drained loam with pH around 4.5 and 7.0. Carambola is typically cultivated from grain but furnace grafting onto the required rootstock contributes to the highest industrial manufacturing process. Just a few days old seeds are effective, and they must be supple and fully established for spreading. Seeds must be cultivated in well-drained potters like the peat moose and ripen, based on the weather within roughly seven days. The soil must be transplanted to containers comprising sandy loam soil when the seeds successfully germinate. The seedlings must be stored before they are transplanted in to such containers. Indirect sunlight and separated by 9 meters (30 ft) apart, Carambian trees should be planted.

For optimal development, carambola trees need soil moisture. After plantation, new trees should periodically be irrigated to facilitate settlement on the land. From planting to fruit set, older plants must be periodically watered. A mulch coating surrounding the trees helps to retain soil but must not be mounted against the trunks. Young carambolas profit from potassium, phosphorus, magnesium and nitrogen fertilizer treatments every 30 to 60 days. Fertilization of mature trees 4 to 6 times per year must be created. For the initial 1–2 years of cultivation, small trees must be sliced by slicing off branches larger than 60–90 cm (2–3 ft) long for branching purposes. To retain a reasonable height, overgrown bushes must be trimmed (Dagar, 2014). Carambola Fruit may not grow from a tree. Just before completely mature it must be collected. Whenever the fruits switch from green to yellow, the tops of the buds stay green, they will be ready to be harvested. Generally, fully mature fruits drop off the tree but if preferred can be selected sooner.

1.8 KARONDA

Karonda fruit is an ovoid shape to oval with rounded ends, with tiny berries, which are on average 1 to 3cm long. Karonda cultivates in clusters of three to ten fruit and matures at various points granting the shrub a multicolored look. When younger, the fruits become green to white, develop into a luminous red and pink color and gradually grow into a dark purple, almost black tone. The thin yet hard skin often gets slightly wrinkled from close, smooth and shiny when it is ready to be harvested.

Figure 1.5: Karonda fruit is a popular summer commodity in India, which is commonly used in picking, jams and preserving products and is known as Black Currant and the Christ Thorn, which is technically known as *Carissa carandas*. This delicious abundance also provides major medical advantages such as heart disease prevention, digestive pain relief and fever healing.

Source: *https://www.netmeds.com/health-library/post/karonda-tremendous-health-benefits-of-the-black-currant-fruit*

Based on the types of maturity the flesh is aqueous and smooth with 2 to 8 flat brown seeds, from pale red through violence. If picked, the fruits will produce a milky white latex until consumption. The fruit of Karonda has a mildly sweet grassy flavor with notable acidic, tasty, bitter, and sour notes. Based on the increasing climate and maturity, every fruit can differ in taste. These fruits can be found through early fall in the summer. Karonda fruits are a small, sour fruit that grows on a woody bush that belongs to the Apochinae tribe, and botanically known as *Carissa carandas*. Prehistoric species are naturalized, mainly as a living barrier, therapeutic element, and nutritious food source, in subtropical and tropical areas across Southern-eastern Asia, Asia, and Africa. Plant species are also available, with some of Malaysia's regions finding karonda to be rare, while the plant is considered abundant in other areas of Africa and India.

Some of the regional names of Karonda fruit include Caranda fruit, the Bengal Currant, Kerenda, Karanda, Nam Phrom and Namdaeng in Thailand, Samarinda, Buah Renda, and Kerendang in Indonesia, and finally Karanda and Keeenda in Malaysia (Dadabhau et al., 2013). There is no commercial production of Karonda fruits. They forage from gardens and wild plants for local market sales. Karonda fruits are great sources of iron for better transport blood oxygen and vitamin C for immunity strengthening, stimulating the development of collagen and reducing inflammation. Calcium, vitamin A and phosphor are also found in the fruits. Karonda's fruits are used in ancient African and Asian medicines to aid with anemia, fever, cold and indigestion due to their non-inflammatory characteristics. Karonda fruit is ideally suited for cooking purposes, as it is neutralized with sugar substitutes. The produce can be consumed crude, but certain berries get a very slick, nearly unappealing taste. Fresh fruit is eaten with chile powder, salt, and sugar, salt in Thailand to complement the savory tastes. The fruits are often placed in salted water to decrease the acidic notes present in the flesh. Karonda fruit is compressed and exudes a brightly round juice, allowing for cocktails, smoothies, and juices an appealing color.

The juice could also be cooked in syrup with fruit spikes, fizzy waters and sodas as well. Karonda fruits are baked as jams and jellies, as well as juices, or filled with cooked products such as pastries and tarts. The fruits can also be mixed in flavorful dishes even curries, and marinated as a condiment when small. Karonda fruits are well-paired with spices including mustard seeds, chili powder, cloves, turmeric vanilla and curry powder, and aromatic products such as garlic, ginger and onióne. The fruits start to shrink as freshness decreases. In the 19th century, the karonda fruit was distributed throughout India as they were embedded in a live fence called India's Great Hedge. The British Empire built the wall to banning the illicit trading of salts across the nation. The Great Hedge took more than 30 years, and the Inland Customs Line which was a British frontier with salt taxation in areas of India which did not have a widely accessible fragrance, was cultivated alongside prickly shrubs, like Karonda. Shrubs of Karonda can develop in height to four meters, forming thick thorny thickets and forming a wall that is impenetrable and strong (Davis et al., 2014). The weeds have also been mixed with decaying branches among certain sparkling foliage, thus creating a barrier that cannot be quickly demolished. The wall has finished being dismantled as the British Reich collapsed and nothing exists of the barrier in present days given the laborious effort to create the Great Hedge. When occupied by the wall, the barren land was converted into a set of streets, but along those same streets, some Karonda shrubs could still be considered a slight reminder of the closed area.

1.9 LOQUAT

Possibly one of the best known exotic fruits crops in Texas is Loquat (Japanese plum or Japanese medlar), but only a limited number of citizens outside of southern Texas are few in the way. It is highly cold and resistant and is mostly cultivated as a northern Texas ornamental in the valleys. The loquat tree is originally from China and has thick, steep leaves. The plant is quite linear and grows individually at the open with a small, thick crown, with a peak of 25 feet with a width of between 15 and 20 feet. Over them, the leaves are vivid, deep green, and underneath are white to rolling. Such tree features have rendered the loquat an outstanding example or emphasis in the home setting.

Figure 1.6: Little, round or pear-molded loquats (*Eriobotrya japonica*) are the trees that only occasionally have the length of multiple inches (5 cm.). The succulent tissue, white, yellow or orange, might be sweet or marginally acidic in taste with orange or yellow stripping. It is delightful when stripped and eaten new, or for later use, you can freeze the entire organic product. They produce amazing pies, jam or jellies.

Source: *https://www.gardeningknowhow.com/edible/fruits/loquat/growing-loquat-fruit.htm*

The matured Locus tree can tolerate 10 °C before severe injury, but at temperatures under approximately 27 the bulbs and the fruits are destroyed. Sadly, in early winter and late autumn, the navel orange blooms and should mature in colder months. Its fruit occurs therefore seldom except in southeastern or central South-East Texas or southern Texas after mild winters. Loquat is extremely suitable for almost all soils with sufficient inner irrigation and fairly not salty soil. Soil pH doesn't seem to matter, because the trees in eastern Texas' acid soils and northern, central and southern Texas' alkaline soils thrive just as well (Das & Prakash, 2009). When considering fruit development, loquats on the South-East or the Southside must be cultivated to get as much cold security as possible from the residence itself. Plant it elsewhere in the ideal landscape.

A huge proportion of choices, many of which are produced in South Florida, have become identified over the years. Since the fruit has seldom attained market significance in the U.S., nursery workers prefer to spread the trees instead of as loquat species. In Texas, the majority of loquats are possibly seeded or spread from seeds asexually. The fruit content of the loquats in Texas is therefore extremely variable. Loquat is quickly promulgated from seed but sometimes self-infertile seeds do not seem to pass from seeding. Veneer and shield grafting are also reasonably effective on rootstocks. Air layering is a great approach to spread from an especially fruitful tree.

A loquat tree collected from the nearby nursery will be produced in soilless newspapers. Since soil-free media are away from an interaction with the ground of a planting site, where water, air or roots cannot be transferred easily, the plant cannot necessarily be removed from the pot and placed inside a growing hole - as growth is relatively slow. Any of the material should be extracted from the edges and surface of the root ball to ensure sustainability and rapid growth to reveal some of the roots. This is ideally achieved with a soft flow of water through the garden tube that is approximately one centimeter from the medium surrounding the ball. The external roots are thus positioned in immediate interaction with the soils of the plantation site when they are planted, thus assuring growth and survival, given adequate watering.

Water at the plantation extensively and for the initial week again once every 3 or 4 days. Extend the period for the following few months between irrigation till the plant is fully founded. To encourage irrigation, create a water ring many inches high and dense on the surface of the peak surrounding the newly established tree, a few feet wide. Cover the ring as required with water. The ring will eventually dissolve into the underlying ground when the plant is grown. Only when recent growth starts to fertilize, one should use the similar fertilizer which they have used on lawn grass or the usual ammonium sulfate (21-0-0 0) or whichever specific fertilizer the country extension agents recommend for your field, excluding a fertilizer that includes a weed-killer. In the first year, two cups in the second and three cups in the third year the young tree can obtain approximately one cup, centered on ammonium sulfate (Dixon et al., 2001). The fertilizer must be divided into three or more requirements per year to achieve optimum performance. Spread the water and scatter it over the surface. Since a small tree cannot cope very well with turf-grass and weed, it must be maintained clear of all certain vegetation for an area of 2-3 feet in circumference centric to the tree. With a systemic herbicide, the undesirable vegetation can be destroyed and a dense layer of natural mulch holds it away.

There should be continued management of the tree with respect to weed control, mulching, and water, as during creation. As far as nutrition is concerned, one pound per inch per trunk diameter is appropriate every year, and divided uses are available in March, June and September. Loquat trees may not generally need pruning, as the tree sets its regular form through pruning, given it has enough space to grow. In 2 to 3 years, loquats can start carrying, with a better formed old tree generating 100 pounds of product comfortably. In general, an especially large crop is reduced in fruit size. The floral panicles usually occur at the ends of the sections in late fall. The petals are aromatic but small and not very visible. In late winter the fruit develops before fall. The fruit generally is around 1.5 cm in length and maybe one-inch-wide, pale yellow-orange, pear-shaped, globose at maturity. It has two to three large, dark and smooth brown seeds. It's juicy and strong. There are variations in its flavor from tangy to sweet, depending on the selection in the variety.

1.10 PHALSA

Phalsa is a native fruit and is grown professionally in various countries. It is a member of the Tiliaceae family. Phalsa is said to have some therapeutic properties in Vedic literature. It can develop within ignored conditions and water shortage, under which only several certain plants survive. The leaves are wide, round, 1-1,5 cm long and have a length of 5-18 cm. The flowers are made of cymes, with each flower being youthful with five large sepals

(12 mm) and 2 cm in diameter, and with a smaller petal content (4-5 mm) percent. The mature fruits have a circumference of 5-12 mm, are edible and purple to black. Its fruit is appropriate for the local market (Anand, 1960) due to its limited shelf life, or must be handled promptly following harvest (Desai and Salunkhe, 1984).

In the provinces of Madha Pradesh, Uttar Pradesh, Rajasthan, Haryana and Punjab, it is commercially cultivated.

The crop develops aggressively under a wide range of soils, namely limestone, clay, and fine sand, and generates adequate soils if soil fertility is not extremely bad. The crops are chlorotic and under water-logged conditions display poor growth. So the best for planting is loamy and drained soils (Fakir et al., 2018). The phalsa has a deep-rooted system and hence is advised for productive cultivating of deep fertile plains of loamy consistency clear from alkali patches or layers of kankars.

Figure 1.7: The phalsa fruit is extremely sensitive; it can't be shipped and must be rapidly and locally eaten. Near large cities that offer a ready market, production is limited. It is therefore available for the few weeks in May-June at a high price. It is also cultivated as fillers in mango and bael (*Aegle marmelos*).

Source: *https://www.downtoearth.org.in/news/food/seasonal-relish-57933*

Except for higher altitudes, the phalsa plantations remain successful throughout India. The crop is feeding stuffy and usually exits gradually in places with a mild cold season, and can withstand light frost at the expense of infestations. It thrives well within varying climatic needs and requirements defense against cold freezing. But sufficient sunlight and mild or warm temperatures, sufficient fruit color production and good food quality are important for fruit maturation.

There are 18 genera and 400 species, and only *Grewia asiatica* has marketing significance. They are transmitted in tropical and sub-tropical areas worldwide. The high wild phalsa increasing crops generate fruits of marginal value, that most customers do not appreciate. A favorite for production is the low-growth form of bushy and/or dwarf phalsa plants that produce a strong blend of acid and sugar. No established varieties are available, but local preferences are accessible for various cultivated regions. The dwarf genetic diversity was noted to bear flowers on every node, throughout the axils of the leaves of the latest year's development. However, various flora results were found in the high genotype 3. The type of dwarf was precocious and dehisces with a greater proportion of seed germination than that of the high type. Top flooring and crop residues (10 kg/pit) are used to stock pitches of 60x60 cm x 60 cm x 60 cm. In July-August or February-March Phalsa crops can be cultivated as crops throw their leaves out of the ground. Seeding is cultivated at 2.5 to 3.0 meters, with one-year-old seeds. It is nevertheless well adapted as it finishes fruiting and flowering within 4-5 months, for high-density cultivation. By boosting the growing conditions, the yield of phalsa can be improved (Galán, 2008). For the growing production of crops, two plants may be cultivated at a length of 0.6 meters, while a new pair of crops are cultivated at a range of 3 meters from the very first pair. This approach doubles the quantity of crops per unit area and raises the overall production by 20%-30%.

1.11 MULBERRY

Silkworm (*Bombyx mori*) is the sole feeder that is cultivated throughout various climates, from tropical to temperate. Mulberry leaves are an essential economic element of sericulture because they affect the quality and quantity of leaves generated per unit area directly. Mulberry is a sturdy, deep-rooted tree that is suitable almost in all soil conditions. It is resistant to drought and can grow in semi-arid and arid zones. It flourishes better with 400 mm and higher precipitation. The optimal demand for precipitation is 800 mm annually. The most appropriate for molt growth is a temperature around 20-30 deg C. This is attributable to the detention of bud development and springing up at temperatures under 13 degrees, whereas seed germination above 40 degrees is caused by high evaporation and transpiration. As the mulberry is a deeply embedded plant, soils must be required to provide even greater depths where the root structure is penetrated with an adequate amount of air, water and nutrients. The soil must be fertile with a strong capacity to retain water, textured, friable, dense and active loam to clayey loam. In the global population, there are approximately 68 genus Morus species in Asia. In Kenya, *Morus alba* is the primary species. Certain species do very well during circumstances of rain-fed, whereas some need irrigation. Albeit throughout the colonial period, the names of those species may have arisen in a foreign country, depending upon the location of the mulberry tree in Kenya initially. Grafting, layering, and cutting is primarily the vegetative propagation of mulberry. Stem propagation has become established as the best approach and is suggested for this reason in the existing research Seed propagation is utilized for growing purposes only. To release the ground, it must be ready by plowing thoroughly up to a depth of 30-40 cm. This operation must be carried out before the storm. Perennial weeds must be banned, such as rhizomes and couch grass. On steep terrain, ideally as one row, more than 30% inclined terraces should be built.

The entire chopping distance is put into the earth, allowing just one bud outside the ground line. Cuts are placed at the nursery 6 cm x 6 cm (15 cm x 15 cm). To maintain the soil moisture irrigation must be performed periodically. This boosts root growth more quickly. A few shoots grow from one cut, but then one can grow. All other shoot are then cut off. When found, weeds are eliminated to decrease food and humidity demand. This means that saplings are moved from the nursery to the main location. It can be transplanted into saplings that are 80-90 days old. During the root removal, the soil structure must not be harmed. The nursery must be well until it is uprooted. The transplantation is performed earlier in the cool weather, in the late morning late or in the night at any time of the day (Gera et al., 2013). The shooting must be minimized by transpiration to minimize loss of water while the roots must be cuts to prevent the sapling from bent roots.

Figure 1.8: Mulberries have lobed leaves, which are serrated on the edge. Individuals can be monoecious (both male and female) or dioecious (bearing just male or female blossoms).

Source: *https://amp.theguardian.com/food/2020/sep/19/the-sweeter-the-juice-an-entirely-ripened-mulberry-is-the-flavour-of-self-control*

The regular culture can be used to plant legumes among the mulberry rows. Even though it is not appropriate to intercrop climbers or other legumes that need pesticide spraying which is hazardous to the earthworms. Cultivation of legumes can be achieved during the initial year when the plant is small, which in the initial year contributes to extra nutrients. Three months on throughout the major fields of the mulberry, trim the leaves at the general point Many buds from the bottom can therefore expand. Once such shoots reach a height of 3ft, pick 3 powerful shoots after approximately 6 months. Spread the harvesting/pruning board at 1ft off the ground. Upon nursery transplanting, cultivation begins for nine months. On average, in the initial season, each plant makes 1 kg of foliage (Gururaj et al., 2017). When ripening in the 3rd year, the tree must produce 2kg of leaf mulberry for each tree, totaling 20m/t/ha/season. 3-5 crops can be produced in a year based on the precipitation patterns.

1.12 WEST INDIAN CHERRY

The Barbados cherry is a little tree or large bushy shrub with a height of up to 20 feet (6 meters) and an equivalent width with tiny buds and a tiny, more or less upright or stretching trunk with a length of up to 10 inches (4 inches) in diameter. Its evergreen leaves are a little wavy, broadly oblanceolate or obovate, oblong, and elliptic, 3/4 or 2 3/4 in (2-7 cm) lengths, 3/8 or 1 5/8 in the width, rounded or obtuse in the apex at the root, sharp or cuneate in the base of the hair.

The flowers are made of 5 lavender or pink, beaded, spoon-like, petals, in a short or sessile peduncle cyme. The fruit after 2 or 3 years of age is round to oblate, clearly three-lobed but cherry-like; 1/2 to 1 in (1.25-2.5 cm) wide; very juicy, orange, glossy, small, round, shiny, pulp. Two wide and one small fluted wings are the 3 small, rounded seeds, forming three leathery, yellow, "triangled," inedible "stones," typically built to be corrugated (Haque et al., 2018). University of Florida's Agricultural Research and Education Center in Homestead specialists, in 1956, after making preparatory assessments and determinations, chose as predominant and named the 'Florida Sweet', a clone that was watched to have an upright propensity of development, huge natural products, high yield, semi-sweet flavor, apple-like, and thick skin.

For young trees, short exposure to 106 degrees F (-2.22 degrees C) can be considered sub-tropical and tropical. Any decrease in temperature lower than 30o F (-1.11o C) destroy in young plants. It is generally sensitive to both medium and low and can survive long dry times, but cannot yield results till the rainfall has come. The tree is adequately rooted in calcareous clay, marl, and stone. At least 5.5 is the pH needed. Upgrading to 6.5 enhances root growth greatly. Acid soils need added lime to prevent and raise nutritional deficiencies. Tapered to 8 in (20 cm) or even more, lime should be incorporated into the ground.

If a seed is utilized in plants, selections must be made from suitable clones that are not subjected to unfavorable cross-pollination. It must be purified, dry and dusted with fungicides. It is also necessary to note that the seeds in a single fruit grow unequally and that they only germinate adequately if the fruit is mature. Germination values can be as poor as 50% or just 5%. When two to three in (5-7.5 cm) high, seedlings from plantations must be moved to containers.

Modified or cleft crown grafting is possible in the summer but is not common, since raising the tree from sectioning is so much easier. Branches with 2 or 3 holes connected hormonal-treated and placed in the sand or other appropriate means under a continuous or sporadic nebulum root within 60 days will be cut from 1/2 to 6 in (6 to 12.5 mm) and from 8 to 10 in (20–25 cm) thickness. They are transplanted to nursery columns and kept in shade for seven months or a year until they are placed on the ground. A year later, several fruits will be produced but no successful harvest until the third or fourth year can be predicted. For approximately 15 years, the plant will proceed to stick well. Through flowering and maximum fruit ripening, there is just a 22 day- period. Grafting is normally only carried out if clippings of the required clones are limited or if there is a nematode-resistant rootstock to save the desired crop; or if a tree with low-quality fruits is top worked.

1.13 LEMON

The actual citrus trees are 10-20 ft (3-6 m) high and have prominent thorns on the branches. Oblong, elliptical or long ovated, 2 1/2 to 41/2 in (6.25-11.25 cm) lengthy, a fine dent with the narrow edges on the petioles, are the substitute leaves, reddish when old, becoming dark-green overhead, lit-green under is their description. The flowers may be slightly scented alone or may be clustering 2 or more in the axils of the stems. They have reddish buds. The open flowers are 4 or 5 petals 3/4 in (2 cm) long; the top layer is white (within), the under (outside) purplish and the yellow anther-like stamens are 20-40; somewhat united.

Figure 1.9: Citrus fruit is common; which individuals utilize to introduce flavor to foods in tiny amounts. But because of their extreme, sour taste, they seldom eat them alone.

Source: *https://www.medicalnewstoday.com/articles/283476*

The fruit is oblong with an apex-like protruding; 2 3/4 to 4 3/4 in length (7 -12 cm). While a few lemons have longitudinal white or yellow and green streaks, the pulp is generally light it is aromatic and has a range of oil secretions 1/4 to 3/8 in thickness (6-10 mm); the pulp is pale□yellow, juicy and acidic in 8 to 10 segments. There are some fruits that do not contain any seeds; most are a few seeds, either elliptical (ova) or ovate, sharp, smooth (9.5 mm).

As the lemon grows more or less continuously, it is increasingly vulnerable to frost than orange and far less capable of recovering from cold wounds. The tree is defoliated to 5.56o-4.44o C, at 22o to 24o F. A decrease in temperatures to 20o F (-6,67o C) causes significant wood destruction until the weather has slowed to almost two weeks. Flowers and juveniles are killed at 29° F (-1.67° C) and almost mature fruits under 28o F (-2.22° C) are seriously harmed. On the other hand, in coastal regions, the lemon achieves the highest production, with summer weather too cold to mature oranges and grapefruit properly. The variety of the lemon is thus comparatively small. Lemons are economically grown in Florida on the West coast of Ruskin and as far north as Ft. Pierce on The East coast (Haque et al., 2018). As a doorpost, the 'Meyer' lemon could be produced anywhere oranges flourish, even to the west as far as Pensacola. The fruits are damaged and the conditions are ready to defoliate the tree and the wind is covered. The citrus fruits are cultivated in both humid and dry conditions, which are primarily uncomfortable in the treatment and storage processes. The yearly rainfall ranges between 25 and

125 cm over a large lemon-growing area in California. The lemon has to be watered in long, dry stretches.

The lemon tree does have a relatively infertile, extremely bad ground history. Haines are primarily on sand in Florida. In California, the muddy clay loam with high water retention capacity has continued to expand very well. Sandy deep, clay, and sand, with porous structure and proper drainage are preferred soils in Guatemala. Black soil is also sufficient if it is not sitting on calcareous subsoils. pH should be around 5.5 and 6.5. For its acidic nature, lime needs to be added to the optimal level. The raw citrus fruit is commonly cultivated. The lemon 'Meyer' can easily be replicated when broad slices are rooting in the nursery and immediately planted in the grooves. The fruit yields 2 to 3 years earlier from the sprouted trees and they have a prolonged life that has been fully developed for more than 30 years.

Company citrus fruit has been budded in Florida to mandarin rootstocks 'Cleopatra,' and 'rough lemon,' sweet orange. The use of the sour orange, Alemow (*C. macrophylla* Wester, old Philippine lemon combination), and Volkamer lemon (*C. volkameriana*) are more current. This is used in soils having excessive dissolved salts and boron in California. The bud union pit kills the tree if it is utilized as a rootstock for 'Eureka.'

1.14 RAMBUTAN

Rambutan is stable in weather ranging from 22–30°C to warmer tropical climates and susceptible to environments under 10°C. It is planted for the market at 12 to 15 degrees above the equator. The tree-line is suitable for growing up to 500 m above sea level and is ideally suited to organically enriched sandy loam, clay loam, and deep soil. It grows on hills, where proper drainage takes place. Budding, air layering, and grafting propagate rambutan. After 2 to 3 years, the sprouted trees will fruit, with optimal performance after 8 to 10 years. Seed bearing trees are planted after 5 to 6 years.

Aril in some commercial varieties is attached to the seed but 'freestone' varieties are available and are in great demand. In general, only one light brown grain is discovered, which would be useful to production and utilized in cooking and the production of soap, abundant in some oils and fats (primarily oleic acid and arachidic acid). In traditional medicine and in the development of coloring, routine roots, leaves and barks have different applications.

Some regions may have rambutan trees bearing fruit twice a year, once within late autumn and early winter, whereas early summer and late spring are less seasonal. Certain places like Costa Rica get a single fruit season, stimulated by budding with the beginning of the rainy season during April, and in September and August the fruit is usually ripe (Hoang, 2008). The fruit has to mature on the tree and is collected over four to seven weeks. The fresh fruit is readily browned and shelf-life is minimal. A standard tree can generate five thousand–six thousand fruits or more (60–70 kg or 130–155 lb per tree).

Returns start on new orchard trees at 1.2 tons per hectare (0.5 tons/acre) and on matured trees can be as high as 20 tons/acre. In Hawaii, in 1997 there were 24 out of 38 grown hectares (60 out of 95 acres), with 120 tons of fruit being collected. Improved crop maintenance, particularly pollination, will boost production and high return compact crops can be planted.

Figure 1.10: The fruit has a similar look when pelted with lychees and long fruits. The sweet but creamy taste and seeds in its midst are its translucent white shell. Rambutan is highly nutritious and can provide medical benefits that range from initial weight loss and improved metabolism to increasing disease resistance.

Source: *https://www.healthline.com/nutrition/rambutan*

1.15 FIGS

Figs are not difficult to fill in warm environments yet produce their best organic product in Mediterranean environments with cool, wet winters and blistering, dry summers. Find out about fig natural science, assortments, developing procedures, and sickness and bug issues (Hossain et al., 2017). Even though figs are a subtropical plant category, developed fig trees are completely strong to 15 or 20 degrees F (chilly temperatures). Individuals who need to develop figs outside the ordinary temperature range should plant them in holders or go to impressive endeavors to secure them throughout the colder time of year.

In the soil, fig plants can rapidly arrive at 15 to 32 feet in stature. The shade can spread wide in a similar manner. The root framework is normally shallow without a taproot and can undoubtedly spread to multiple times the width of the overhang. Preferably, fig plants ought to be planted in very much depleted topsoil with a lot of natural issues, yet they will endure normal to helpless soil. Whenever they are set up, they are to some degree dry season lenient, likely as a result of their extremely broad and wide-running root framework. Figs endure soils with a pH going from 5.5 to 8.0. Cultivators who have soils that are acidic need to apply lime to raise the dirt pH up to the favored value of the fig, a pH of 6.0 to 6.7.

Shooting growth below 18 inches is indicative of the need to add fertilizer. In springtime, mid-May and mid-July, applying balanced feedstocks like 8-8-8 for plants 1-2 years old at the pace of 8 ounces for each plant. Good seedlings may be fertilized once in Early Spring if necessary. At growing, no fertilizer can be used. Implement fertilizer 1 feet from the crown across the base of the plant. Figs adapt to strong compost and manure treatments quite well (greater than many fruit trees). Make sure you do not use fertilizers too far in the growth period as it stimulates new shoots that cannot stiffen until winter. A full long-acting fertilizer, like Osmocote plus micronutrients, is a good option when cultivating figs in pots. Use large quantities of manure and high-nitrogen fertilizers like alfalfa, soybean or cotton plants to grow organically.

1.16 KIWI

Kiwi is mainly cultivated in Kerala, Arunachal Pradesh, Meghalaya, Sikkim, J&K and Uttar Pradesh. Forecasts of the area under cultivation have not yet become implemented until recently in the region. The fruit is nutritionally and medicinally beneficial. It is a rich vitamin C and B source and minerals such as calcium, potassium, and phosphorus. In deserts and salads, the fruit is eaten fresh or mixed with other fruits. It is also utilized in wine and squash preparations.

In areas between 700-800 chilling hours, Kiwi can be filled (no. of hours the temperature stays at or under 7oC through the colder time of year season). The plant is 800-1500 m above m.s.l away. recipitation of 150 cm is appropriate /year. During the construction timeframe, the precipitation should be all over. Due to its incredible vegetative growth, enormous leaf and wine inclination, the plant cannot withstand solid breezes and ice during its development. High temperatures (> 350 C) and low humidity can cause leaves to burn in summer. The key problems of its growth on the lower territories are sunshine and warmth tension. In some areas of Himachal Pradesh and parts of Kerala, the plant can occur. These two states can grow the crop commercially (Islam et al., 2019). Tomuri, Monty, Hayward, Bruno, Alloson and Abbott are significant kiwi varieties grown in India. The rolling land for cultivating vines is divided into terraces. The rows should be north-south-oriented to allow the best possible use of illumination. For the effective production of viticulture, comprehensive preparation of the soil is necessary. Pit filling, farm manure mixture, and pits shall be finalized by December. Plants are primarily asexually propagated by grafting and cuttings. In January, seedlings are normally finished. The plantation should be carried out at the same profundity as the crops in the nursery. The soil needs to be deeply rooted. The crops are cut to approximately 30 cm. Encouraging energetic development. The seeding distance depends on the variety and training method. Pergola and T-bar planting methods are typically used. In T-bar, 4 m apart; 5-6 m from row to row. The plant is specific from plant to plant, whereas a 6m distance is typical with pergola. It should be held from row to row.

Figure 1.11: While not juicy, the fig is a crop with an unbelievably luscious fragrance and a pleasant taste. Initially from Asia, figs now thrive in the Mediterranean, and thousands of hybrids are divided into four principal colors: black/ purple, red, and white.

Source: *https://www.bbcgoodfood.com/glossary/fig-glossary/amp*

A manure portion of 20 kg. yard excrement (basal portion), 0.5 kg. NPK blend containing 15% N is suggested for application consistently. Following 5 years old, 850-900 g. N , 500-600 g. P , 800-900 g. K and barnyard fertilizer ought to be applied each year. Kiwi requires high Cl since its insufficiency antagonistically influences the development of shoot and roots. Conversely, abundance levels of B and Na are destructive (Jamnadass et al., 2011). The N manure ought to be applied in two equivalent portions, half to 66% in January-February and the rest after organic product set in April-May. In young plants, the manure is blended in the soil inside the borders of the plant, and for the developed plant it is applied equally over the whole soil surface.

1.17 BREADFRUIT

Artocarpus altilis was brought into the world in the archipelago of Malaya and was regarded in 1788 for being a piece of the Bounty, the celebrated boat of Captain Bligh. There were a great many breadfruit trees bound for the Western Indian islands locally available the Bounty. The natural product is delivered in the United States in South Florida or imported from West Indies, particularly Jamaica, regularly from June to October, all year and can be found on neighborhood forte business sectors.

The breadfruit tree arrives at an elevation of around 26 m and is covered with huge, thick and thick leaves. A smooth juice called latex, which is particularly valuable for some, things like boat caulking, is delivered in the whole tree. The plants are blossoms for the two people that develop on similar trees (monoecious). Male blossoms initially arise and, a couple of days after the fact, female sprouts are pollinated. The subsequent organic product is round to oval, 6-8 cm long and approximately 8 cm (20 cm) in weight.

The skin is green and dainty and in the long run develops into a light green for certain rosy earthy-colored regions with unpredictable polygons. In development, the organic product is white inside, boring. The organic product is hard and boring as a potato, when green or under develop. Breadfruit is principally utilized as a vegetable and has a musky fruity taste when cooked, yet amazingly gentle, and is generally excellent for trying dishes like curries. It has been cooked. Ready natural products may have a ready avocado surface or runny brie cheddar. Breadfruit is one of the world's biggest food preparing plants. Up to 200 or significantly more organic products can be created per season by a solitary tree. Profitability shifts relying upon the field of wet or dry cultivating. The organic product is wealthy in potassium. It very well may be cooked, steamed, prepared or seared in close relationship with a potato. Eliminate the white, dull sap or latex with the bread natural product for around 30 minutes before use. Another entrancing reality is that this is firmly connected with the "jackfruit" and the "bread-nut."

This species can ordinarily be found under 2,130 meters (650 meters) rises, however can be seen up to 5,090 feet in tallness (1550 m.). It fills in sandy soil, cream, sandy topsoil or sand in nonpartisan to basic soil. It endures even salt grounds. Root cuttings and air-layered plants were shipped over immense sea distances by the Polynesian people groups. The organic product was a significant source of sustenance, yet in addition to structures and kayaks, the lightweight, safe wood. as the latex was used as a pulverizer and to catch birds . The wood pulp is used for paper and even utilized therapeutically (Kotwal et al., 2011). The customary staple of the Hawaiian public, poi, produced using taro root, can likewise be supplanted or added to bread. The subsequent natural product poi is known as poi ulu.

As of late, researchers have discovered the three best mixes or soaked acids for repulsing mosquitoes (lauric, undecanoic, and capric acid). Nonetheless, we keep on finding new applications for this incredibly adaptable plant thinking about the chronicled and social estimation of the breadfruit.

Chapter 2

ROLE OF MINOR FRUITS IN FOOD AND LIVELIHOOD SECURITY

CONTENTS

2.1 INTRODUCTION

The world is currently at risk of population growth without the ability to sustain itself. Researchers claim that by the year 2050, the population may stabilize at about 9.2 billion people. Considering the behavior of human beings in the world, agricultural and natural resources have been degrading or rather fast-shrinking over the years. In the current society, technology development is what the majority of the countries are after (Akinnifesi et al., 2002). Such firms require large sizes of land to be built in. Furthermore, from technological industries, there are high emissions of polluting substances that are harmful to both the soil and air. This has led to the negative effects of soil and effects on the ozone layer meant to protect us from the heat and rays of the sun. On the other hand, most firms/industries always emit their wastes to the rivers and natural forests. This has made the majority of the water from the river undrinkable and made a lot of trees die because of pollution (Koohafkan & Altierim, 2011). It is claimed that, if human beings are not careful by ensuring that they make the world a better place, then it will no longer be a safe place for us to live in.

Figure 2.1: Pollution by plants is the major cause of environmental degradation which will be most harmful to future generations.

Source: *NPR*

The fact of the matter is that the effects of our actions are already visible. Some of the countries have a very small number of the elderly represented in their population. Also, there is the melting of ice in the arctic which is causing the water levels to rise. With the rise in water levels, some of the countries that are not that far above the sea level will submerge in water. This is something that needs to be addressed since the fate of the future generations depends on the current generation's act of responsibility. One tree a day will make the world a better place to live in. If trees a planted, they may help in absorbing the polluted air and further making the soil firm enough to avoid erosion. It should be understood that technological advancement is beneficial to society but it should not be at the expense of killing all living organisms including the human population (Ahmad et al., 2008). Currently, because of such degradations, the production of crops is at high risk. For instance, researchers claim that, by the year 2025, about 30% of crop production will be at a high risk because of the reduction in viable water.

The world is therefore left at something that seems like a confused state. The population is increasing while on the other hand, the future of crop production is at a high risk. For us to be able to meet the high demand for food and primary production, we need to be able to produce more and more from less and less size of land or rather the finite natural and non-renewable resources. Fruits, for instance, are both food and cash crops that are deemed as beneficial to the human body. These types of crops are majorly produced in India, which is blessed with many varieties of the same. The country is filled with a large variety of tropical and sub-tropical, minor and underutilized fruits. Minor fruits do not necessarily imply that these fruits have limited value to the human body. Ideally, all fruits are important to the human body and should be regularly consumed. The definition of minor fruits is that they are those that do not occupy a large area to justify their economy of production.

Because of this fact, the majority of the human population should be well conversant with the minor fruits. This is because, these fruits can be easily grown in the home gardens, orchards and are in most cases found in the natural forests. This means that they are easy to grow and are also convenient. It does not require any skill but rather a hoe for the individual to prepare the land for planting this variety of food products. These fruits are said to have both food and financial values and can help in the production of many delicacies such as jam, jelly, pickle, juice and squash. In India for example, people value fruits to an extent that May, June and July are treated as fruits festival months. At these times of the year, the majority of fruits are mature and available and ready to be eaten. At the above stated months, the Indians even go to an extent of praying to their gods to thank them for the ability of their land to give way for the fruits to ripen and ready to eat.

The majority of the fruits that are produced in India are eaten by the local population of which the surplus is what is exported to other countries. This shows how much the individuals in his country cherish fruit production (Altendorf, 2018). Apart from May, June and July, some varieties of minor fruits are available throughout the year. The number of minor fruit tree varieties that produce fruits throughout the year is limited. On the other hand, the availability of most of the minor fruits that are found in India is said to be highly seasonal. This means that these fruits can only be produced once in a single year.

The summer or rather the rainy season is the season in which the majority of the minor fruits are produced. These are the months of May, June, July and August. These are the months when the litchi, mango, pineapple, guava and jackfruits flood the markets. This shows that the availability of minor fruits in a certain country is important. This is because such fruits offer a great opportunity for the members of the public in the particular country in terms of securing their incomes. Additionally, it ensures that there is the provision of food to the vast majority and therefore food and nutrition security.

Figure 2.2: Majority of the minor fruits are produced during the summer seasons in the months of May, June, July and August.

Source: *Maharishi Ayurveda*

The sad truth is that in the world, many of the minor fruits remain to be underexploited. This is majorly due to the lack of awareness and also due to ignorance. Individuals are not involving themselves in the production of such fruits since they do not know the fruits' potential, their demand in the market and have a low and erratic bearing in most instances. There are over 3000 tropical fruit species that are either wild or domesticated or in other instances both. Some of such fruits are cultivated on large scale. The minor fruits are better known for areas of diversity but have not yet been utilized to their full potential. Although, there is a diversity of fruits in the world, some of the species that have gained popularity include mango, guava, citrus and banana fruits. In the Indian system, the local members of the society describe such fruits as minor ones and are used also for medicinal purposes (Banyal et al., 2019). If these fruits are harnessed properly, their potential as a source of food other than meeting the multipurpose needs of the local communities will be realized.

The positive effects realized by individuals from consuming minor fruits tend to be important both in the local communities and also in a particular region. However, there are always negative features of these crops. They mainly involve geographic restrictions. The minor fruits or rather fruits in general have certain biological limitations that may prove to be difficult when dealing with its cultivation. Looking at India for example, the land seems like a desert and rain is just a limited amount. The other reasons in which geographic restriction is considered to be a limitation include the fact that there is limited dissemination of knowledge regarding how to cultivate the minor fruits and the fact that individuals are not educated on germplasm. Apart from geographic restrictions, there are also other negative features of the minor fruits. Some of the species are limited by some other inherent biological factors including the fact that these fruits may prove difficult to cultivate, some of them tend to be low-yielding fruits, are always at risk of contracting some diseases, may prove to have a low storage capacity.

Fruits are generally perishable but after harvest, one can store them properly and be able to use them for longer periods. In other instances, propagation is difficult especially when it is difficult for the farmer to distinguish between the male and the female species or when the farmer has only a single sexed farm. When there are problems in propagation, the difficulty is not only on distribution but also they tend to restrict individuals from manipulating the breeding programs. The fruit species tend to have numerous uses and as such, some of the techniques of crop improvement that tend to focus on only one fruit character are considered to be inappropriate (Barua et al., 2019). This shows the fact that minor fruits are important to the human body but they need proper care from the farmers involved in their production. However, they are not extensively cultivated in the majority of the regions. On the other hand, its consumption and trading tend to be more limited when considering the geographical and quantitative factors although most of them are considered as important in the regional market of areas they are cultivated. Some of the fruits that are said to be minor, can be known as major fruits in their region of influence. This shows that depending on the region that the fruits are located, minor fruits can also be recognized as major fruits.

The minor fruits can be very useful to society and can make a major contribution to the nutritional status of individuals. They are considered to constitute a major diet in various local communities in the tropics. Most of the minor species of fruits are generally adapted to the marginal lands and require minimal amounts of inputs, and thus considered to be very helpful in terms of ensuring the survival of some of the poor communities while ensuring that there is the sustainability of agricultural ecosystems. It is with no doubt that minor fruits are low yielding, the fact is that, for one reason or another, and they are an ignored species of crops. All stakeholders involved in the field ranging from, extensionists, policymakers, decision-makers, donors, technology providers to researchers have ignored these species which has made them vulnerable in society.

Figure 2.3: Bananas are minor fruits that are quite easy to cultivate and require minimal input and effort. they can be grown at the home garden which makes them very useful, especially to resource-poor individuals.

Source: *Britannica*

Apart from the above-stated stakeholders, consumers have also ignored the existence of such fruit species. The world is changing and therefore, the majority of food crops need to be researched on so as to ensure that they can be able to adapt to the changing environmental conditions for generations to come. For the case of consumers, the absence of a certain fruit in the market or even in your firm in places where they used to be must raise an alarm in the household with regard to the reason behind the absence of the same. The consumers may decide to take part in the same to ensure that such fruits are protected by all farm stakeholders in society (Bisht, 2020). As such, not much is known with regards to the agronomic features of the fruits, how a cultivator is required to select the species, what needs to be done to improve the yield of the fruits, some of the practices that need to be done after the fruits have been harvested in preparation for flowering, some of the commercialization processes and the potential markets for such fruits. The lack of knowledge on the above facts shows how little or even no research on the minor fruits are done. This acts as a challenge to the field actors since this plant species has a high potential of saving the human world. As seen before, minor fruits involve those fruits that can be easily grown in the garden with potential for commercial development. These fruits also have a potential for nutraceutical purposes that include their medicinal and cosmetic uses. Some of them also play a major role in controlling pests or rather as natural growth regulators. Other uses of minor fruits include, using them as timber, as fuel, oils, waxes for industry, gums, used as livestock feeds and also have an intrinsic value where they are considered to be genetic resources.

2.2 DIVERSITY OF MINOR FRUITS

Arunachal Pradesh, Assam, Meghalaya, Manipur, Tripura, Mizoram, Nagaland and Sikkim are eight states that are found in Northeast parts of India. The area covers about 7.7% of the total Indian geographic area which is claimed to be the supporter of about 50% of India's biodiversity, whereby, endemic is said to constitute 31.58%. The above-stated areas are considered to be one of the hottest spots in the country and are known for their diversity, diverse nature of the soil, topography and climatic conditions (Behera et al., 2016). The region is known for its diverse species of fruits ranging from *Gynocardia odorata, Mangifera sylvatica, Myrica fraquhariana, Myrica nagi, Pyrus pashia, Prunus nepaulensis, Prunus jenkinsii, Passiflora edulis, Paramaria polyneura, Phyllanthus acidus, Physalis peruviana, Rubus ellipticus, Rhus semialata, Spondias Mangifera, Syzygium jambose, Castanopsis indica, Calamus erectus, Carallia lucida, Dillenia indica, Docynia hookeriana, Eugania cymosa, Elaeagnus pyriformis, Elaegnus latifolia, Emblica Officinalis, Flacourtia jangomas, Averrhoa carambola,*

Artocarpus lakoocha, Baccaurea sapida, Carassia carandas, Chrysophyllum roxburghi among various other species. In the evergreen forests of Assam, Khasi, Jantia hills of northeast India, the known species of *Garcinia* grow extensively in a semi-wild state. For culinary purposes and folk medicine, the members of the region sundry the slices of fruits. Some of the species such as the *Garcinia pedunculata, G. kydia, G. cowa and G. lancifolia* are considered to be some of the most important species in the region of northeast India.

Figure 2.4: The Garcinia species grow extensively in a semi-wild state in India and are considered important in the northeast regions.

Source: *ResearchGate*

2.3 MINOR FRUITS FOR NUTRITIONAL SECURITY

The minor fruits in the world are known for their nutritional benefits. They are considered to be important sources of vitamins such as pyridoxine (B_6), riboflavin (B_2), carotenoids (A), thiamine (B_1), ascorbic acid (C), niacin (B_3), and folacin, fat, dietary fiber and protein. Other nutrients that are provided by fruits include minerals such as, phosphorous, zinc, potassium and calcium (Boyd et al., 2007). Some of the nuts that are grown in various parts of the world are considered to be minor fruits. Nuts are considered to be essential in the body and play a crucial role in bodybuilding. They are also considered to be essential providers of fiber, fatty acid, minerals and also a good source of Vitamin E.

The importance of most of the food resources including fruits are numerous but due to certain reasons, they have not been utilized. Furthermore, their potential to be produced in certain areas is huge but scientists or rather researchers of the same have not been able to realize them. For this reason, the majority of the resource-poor individuals that tend to reside in the rural areas do not have the luxury to benefit from such a considerable advantage of having abundantly available resources. The solution is rather closer than we think and also easier than we deem it to be. It lies behind value addition, educating villagers on the profitable ways of cultivating the fruits and of their nutritional value and finally, providing them with markets. Fruits are food crops that are always in high demand with people in the urban centers having knowledge of their nutritional and health benefits. Indulging in the processing of the minor fruits and value addition, these minor fruits may prove to be useful, digestive and convenient products (Chakraborty et al., 2011). When this is done, the economic value of the minor fruits will be realized and in the process ensure that there is nutritional security for the villagers dwelling in the area in which the fruits are planted.

Figure 2.5: Resource-poor individuals in the rural areas need to be educated on the production of minor fruits in order to be able to improve health-wise and economically.

Source: *Twitter*

2.3.1 Sources of Hydroxycitric Acid

Hydroxycitric acid is a natural diet ingredient that is present in *garcinia sp.* Of fruits. It is said to be an anti-obesity compound that is that are found in the leaves and fruit rind of the *garcinia.* The claim is that, this compound inhibits the synthesis of lipids and fatty acid in the systems of most living organisms. HCA is also considered to be a hypocholesterolemic agent. When measuring different compounds in fruits in terms of weight, HCA constitutes about 20-30%of the fruit.

2.3.2 Sources of proteins

Protein is very important to the human body, considering the fact that toddlers live on breastmilk, which is 70% protein before they are weaned by their mothers. They constitute the major tissues and cells in the body and are said to be the major components of muscles, body fluids like blood and other tissues that are considered to be vital in the body. Those proteins in the human body that form the hormones and enzymes are crucial as they are involved with a wide range of metabolic processes in the body. The wear and tear that occurs in the body are taken care of by proteins as they supply the bodybuilding materials in the body system. One vital role played by the proteins in our bodies is that they act as antibodies that help the body to fight against any infections that may occur in our bodies (Chakraborty et al., 2011). As such, it is seen that proteins are very important in our bodies and are adequately required. For instance, male individuals require about 60g of proteins while females require about 50g of proteins in their daily diet.

2.3.3 Sources of Fat

Fat is something not favored in society because the majority of the human population is composed of women. Fat is an enemy to most women especially those who are trying to maintain better body shapes. But they should not be worried. Manufactured fat is what most scientists consider to be harmful. However natural fats that are found by eating fruits such as the avocado are very useful to the human body (Challies & Murray, 2011). They are considered to be a very important component of the diet and are considered to play numerous roles in the body. In a way, fats are a source of energy but tend to supply per unit energy more than twice supplied by either carbohydrates or protein. It is seen to bring out the sweet flavor of food or rather a diet and tends to slow down the process of stomach-emptying. In the body of a human being, the fatty acid that is considered to be essential to the human body is at 3-6% of the total energy that is ingested.

Figure 2.6: Avocado fruit is a good source of fat that is not harmful to the human body but can be very useful.

Source: *NDTVFood*

There is also the consideration of other variables such as age, gender and the psychological state of an individual. As a matter of fact, for a woman who is experiencing the growth of a fetus, pregnancy and lactation, the fatty acid that is essential to the body is required to be at higher levels than a woman not experiencing any of the stated situations. This however does not mean that the woman needs to put a lot of oil into the food that is cooked. To obtain this requirement, visible fat intake should be about 15-25g every single day. The intake should be in terms of the oil content of fruits and other products. In most of the lower-income rural households, fat is considered to be a major nutritional concern. As such, most people in such areas tend to suffer from phrynoderma (toad skin), weakness, and weight loss (Crush & Frayne, 2011). As such, individuals from such regions should be educated on the various ways to cultivate the minor fruits to help them in their diets and even go to an extent of being financially important.

Some of the minor fruits that are rich in crude fat content include; Bael, scientifically known as *Aegle marmelos* contains about 0.2g/100g of crude fat. Karonda, scientifically known as *Carissa carandas* contains about 2.6g/100g of crude fat. Aonla, scientifically known as *Emblica officinalis* contains about 0.1g/100g of crude fat. Jamun scientifically known as *Syzygium cuminii* contains about 0.15g/100g of crude fat. Soh Lyngdkhur scientifically known as *Morus indica* contains about 0.48g/100g of crude fat. Soh Phie, scientifically known as *Myrica nagi* contains about 1.40g/100g of crude fat. Soh Phienam, scientifically known as *Myrica esculenta* contains about 1.36g/100g of crude fat. Sohiong, scientifically known as *Prunus nepalensis* contains about 0.44g/100g of crude fat. Humra guti, scientifically known as *Terminalia belirica* contains about 1.55g/100g of crude fat. Lateku, scientifically known as *Baccaurea sapida* contains about 0.10g/100g of crude fat. Poniol, scientifically known as *Flacourtia jangomas* contains about 0.13g/100g of crude fat. Sohlang, scientifically known as *Eleagnus latifolia* contains about 0.06g/100g of crude fat.

2.3.4 Source of Carbohydrates and Fiber

Carbohydrates are a class of substances that are ingested in the human body which yield energy. The commonly known ones include glucose, starch, sugar and sugar cane. The main source of energy in the body is glucose that is derived from starch and other types of sugars in a particular diet. In the tribal areas of India, the carbohydrates are derived from particular minor fruits and are considered to be the chief source of energy (Chazovachii, 2012). The energy requirement in the human body generally depends on the age of the individual and the work that the individual is subjected to. It is therefore advisable for most individuals to ensure that there is a presence of carbohydrates in their diet. When one is carbohydrate-deficient, many health complications may be observed in them including weight loss, weakness and the reduction of physical activity. Such complications may lead to a person not being able to conduct their daily activities which makes it hard for them to earn income. Some of the fruits that are rich in carbohydrates and fiber content include, Bael, scientifically known as *Aegle marmelos* contains about 31.8g/100g carbohydrate, 137.00Kcal/100g of energy and 3.0g/100g fiber.

Figure 2.7: Bael fruit, commonly known as wood apple fruit has the highest carbohydrate content compared to the majority of minor fruits.

Source: *123RF*

Karonda, scientifically known as *Carissa carandas* contains about 0.5 g/100g carbohydrate 42.00Kcal/100g of energy and 0.6g/100g fiber. Aonla , scientifically known as *Emblica officinalis* contains about 15.0 g/100g carbohydrate 65.00Kcal/100g of energy and 1.9g/100 g fiber. Jamun, scientifically known as *Syzygium cuminii* contains about 14.0g/100g carbohydrate 62.00Kcal/100g of energy and 0.3g/100g fiber. Soh Lyngdkhur, scientifically known as *Morus indica* contains about 8.10 g/100g carbohydrate, 37.12Kcal/100g of energy and 0.10g/100g fiber. Soh Phienam, scientifically known as *Myrica nagi* contains about 21.83 g/100g carbohydrate, 110.52Kcal/100g of energy and 2.15g/100g fiber. Soh Phienam, scientifically known as *Myrica esculenta* contains about 21.55g/100g carbohydrate, 109.04Kcal/100g of energy and 1.44g/100g fiber. Sohiong, scientifically known as *Prunus nepalensis* contains about 26.96g/100g carbohydrate, 189.80Kcal/100g of energy and 1.70g/100g fiber) (Chakraborty et al., 2011). Humra guti, scientifically known as *Terminalia belirica* contains about 34.40 g/100g carbohydrate, 166.79Kcal/100g of energy and 2.14g/100g fiber. Lateku, scientifically known as *Baccaurea sapida* contains about 7.47g/100g carbohydrate, 33.89Kcal/100g of energy and 2.93g/100g fiber. Poniol, scientifically known as *Flacourtia jangomas* contains about 2.00g/100g carbohydrate, 13.57Kcal/100g of energy and 1.63g/100g fiber. Pora amlakhi, scientifically known as *Phyllanthus acidus* contains about 3.11g/100g of carbohydrates. Sohlang, scientifically known as *Eleagnus latifolia* contains about 9.39g/100g carbohydrate, 42.02Kcal/100g of energy and 1.18g/100 g fiber. The above-stated fruits work much wonder in ensuring that there is energy in the body compared to processed carbohydrates such as potato chips, crisps, biscuits among others.

2.3.5 Source of Vitamins

Vitamins are essential to the human body. In a way, vitamins ensure that the body is kept safe and allows for faster and better digestion of different types of food. Mostly when one visits the doctor when they are sick, they are told to take vitamins in their meals (Dagar, 2014). Apart from the above-stated functions, vitamins tend to carry vital functions in the entire body and the majority of them are involved in the utilization of a majority of nutrients including carbohydrates, proteins and fats. Even though nutrients are vital to the human body, they need to be taken in smaller quantities.

2.3.6 Source of Minerals

If one talks about minerals, what comes to mind is that they are mined. However, they are components that are found in food and are also trace metals that are found in the body. In many metabolic reactions, minerals act as catalytic agents while others form parts of body structural components. Most of the bones that are found in our bodies are made up of calcium, phosphorous, iron, magnesium or manganese. They can either in the structural part or activate a large enzyme system. Intake of food products that are rich in minerals ensures that the bones remain strong and effective in terms of their functionality.

2.3.7 Sources of Antioxidants

In our everyday life, the food we consume tends to have some form of defects or rather items that are harmful to the body. In most instances, individuals tend to experience food poisoning. If that is the case, then one is required to take a certain fruit to help in easing up the pain or the upsets. Fruits that are rich in antioxidants help lower the incidence of degenerative diseases such as cancer, arthritis, arteriosclerosis, inflammation, heart disease, brain dysfunction and decelerate the aging process (Dadabhau et al., 2013). The known definition of antioxidants is that they are substances that tend to delay or rather prevent oxidative damage of nucleic acids, lipids and proteins by reactive oxygen species which is inclusive of reactive free radicles such as peroxyl, alkoxyl, superoxide, hydroxyl and some non-radicles which include hydrogen peroxide, hypochlorous among others. The antioxidants tend to scavenge radicals by preventing the initiation and the chain-breaking processor in other instances suppress the formation of free radicles. This is done by binding the metal ions which then reduces the hydrogen peroxide and quench the singlet oxygen and superoxide. Some of the known antioxidants that are found in fruits include Vitamins A, B, C and E and also the carotenoids that are found in small quantities or rather at a lesser extent in some fruits. The majority of the known minor fruits are considered to be sources of compounds that are active and tend to show the process of antioxidant activity (Davis et al., 2014). Phyllemblin, kaempherol, quercetin and rutin are the active compounds that are found in the aonla fruit, marmelosin is the active compound present in bael while carambola is rich in oxalic acids. Zeaxanthin and lutein are active components that are found in Poniol.

Figure 2.8: Most fruits are useful in the prevention and dealing with diseases. Berries for instance, are useful when dealing with arthritis.

Source: *Healthline*

2.3.8 Source of Ayurvedic Medicine

As seen earlier, some of the local communities in India used the minor fruits grown in the areas that are known for their nutritional and health benefits. In science, minor fruits are considered to be a major source of raw materials in making drugs and traditional medicine. It is with no doubt that the products from minor fruits are considered to be of explicit quality and are considered to have great nutritional and health benefits while also being seen to be of great economic importance. Some of the modern ailments known in society such as cancer, jaundice, asthma, diabetes and nutritional deficiencies can be remedied by the use of minor fruits. In India, some of the minor fruits including aonla, bahera, and are used in the manufacture of more than 200 patented drugs in the country. Bael which is seen to contain the majority of nutrients is used in the manufacture of more than 60 patented drugs. There are even fruits that are currently used in the docking of the AIDS virus in the human cells (Das & Prakash, 2009). The fruit is known as the black mulberry. Medical specialists have gone to an extent of asking individuals to include nuts, vegetables and fruits which are known to reduce the risk of contracting some of the serious ailments such as cancer, stroke, heart failure and various other chronic diseases. Some of the fruits are said to be important in acting as detoxification agents and in other circumstances tend to alter the course of growth when dealing with tumor cells.

2.4 TYPES OF EDIBLE MINOR FRUITS AND THEIR NUTRITIONAL VALUES

Minor fruits are commonly referred to as the type of fruits that grow in a scattered and neglected way for example they grow on roadsides, homestead land, and wasted lands. However, internationally minor fruits are still regarded as a niche product. There are only a few varieties available in the markets places either through ethnic markets majorly eyeing the migrant consumers who are mostly of Asian origin, or premium retail channels targeting conspicuous and affluent consumers. Consequently, other market opportunities have emerged with the demand growing rapidly worldwide including China (Dixon et al., 2001). This has made the production of these fruits to be an enterprising venture for many farmers especially with the increasing response of them having been said to be increasing health and nutritional value among mankind. The demand is set on an increasing trajectory in the developed markets, most importantly the US and the European Union.

In the last three decades, the production areas under these fruits have increased rapidly. This is as a result of many farmers seeking the elevation of malnutrition and losses from their farms with the increased threat of climate change thus they have focused their attention to this promising crop. Many of them have been known to be rich in antioxidants and nutraceutical properties. These crops are also very relevant in regards to poverty alleviation in the rural areas and in areas that are hot, arid and with fragile ecosystems. In 2016 and 2017, minor tropical fruits suffered some weather-related declines arriving at an estimated 24 million tonnes. India and China rank as the largest producers of this crop, accounting for 24 percent and 22 percent of total global output in 2015-2017 respectively, but their produce is almost exclusively destined for domestic markets (Dixon et al., 2001). Other significant producers are mostly located in Southeast Asia, with Thailand, Indonesia and Viet Nam accounting for approximately one-third of global output combined. Brazil is the leading producer of passion fruit and has an estimated share of 7 percent of total global minor fruit production in 2015-17. Again, production in Brazil is primarily for local consumption. The available and growing strong domestic demand has helped sustain and boost price levels. A lot of the tropical fruits have nevertheless remained unexploited due to lack of awareness of their existence and potential, lower availability and location specificity. The various diversity of these fruits are large with about 30000 plant species available for human consumption currently, 7000 plant species reported to be used as food in history while about 150 plants are cultivated for commercial purpose and 103 plants solely contribute to more than 90% of the world's source of calorie where less several hundreds of species remain underutilized or less utilized in the wild areas. The majority of these species are even so facing the threat of extinction due to climate change, large-scale urbanization and deforestation. To safeguard the existing diversity of underutilized fruits and to achieve sustainable development based on the use of available health wealth under various ecosystems, promotion and conservation of these species are of great importance (Fakir et al., 2018). Commercial production and processing for value addition of these crops would enhance the income of small and marginal farmers, improve their nutrition and also help in on-farm conservation of important germplasm.

Figure 2.9: Jack (*Artocarpus heterophyllus*) plant and fruits.

Source: *trees Planet-blogger*

Some of the minor plants include the jack (*Artocarpus heterophyllus*), which is mostly grown in the native Western Ghats of South India. It is adaptable to varied types of soils and climate thus it is commonly available in most regions. jack fruits are a rich source of vitamin A and C and they are readily digestible with sugars of high calorific values. The young and mature fruit seeds can be used in food preparations as they are known to produce a very appetizing aroma. Some of the products processed using the jack seeds include chips, pappad, jam, jelly, stach, etc. There are five species and about 26 varieties maintained in the genetic diversity park for detailed analysis. The jack plant is additionally known for quality valuable timber. Medically jack plant leaves help in improving glucose tolerance in persons with diabetes disease. Jack also has powerful laxative and anti-aging properties that are mostly used by humans internationally, especially in the western regions to keep young (Galán, 2008). The fruits are very effective in lowering high blood pressure in hypertension patients and are very essential to the health-conscious individuals in society. This entails especially the old, the children, the pregnant women and the sick people. This is because it contains anti-oxidants that help in enhancing and promoting better lives for this group of people.

Custard Apple (*Annona spp*) is another type of minor fruit with its native in South and Central America. This genus has a large number of species having enormous economic importance. Firstly, three species are popular for edible fruits by humans and animals where several are used to prepare food products such as ice cream, sweets, juice, pastries, chocolates and mixed fruit jam. They are a good source of muricata ,which is used for medicinal purposes of curing cancer patients. The seeds and leaves of the custard apple have insecticidal properties and can be used for the manufacture of bio-pesticides. the nutritional values if consumed by humans include the anticancer agent. This would prevent persons from developing cancerous cells in any part of the human body. The paste of their leaf of the *Annova* genus is however used to treat boils, abscesses and ulcers to the sick. It is also known for curing diarrhea and dysentery (Gera et al., 2013). Magnesium in custard apple is good in the prevention of strokes and heart attacks. Moreover, regular intake of custard apple reduces the chances of rheumatism and other forms of arthritis as it helpful in the flux out of excessive acids. Additionally, custard apple is an excellent source of copper. Its consumption by humans with challenges of the brittleness of bone is highly recommended as the illness is mainly due to lack of enough copper in their bodies.

Figure 2.10: Custard Apple (*Annona* spp) used for medicinal purposes of curing patients

Source: *Wikipedia*

Avocado fruit also falls under the group of minor fruits. This fruit originated mainly from tropical America. The first growing areas are Mexico and Central America. Avocado is among the most nutritive fruits and is regarded as a major contributor to the human diet across the Globe. Its pulp is rich in proteins and fat, but low in carbohydrates. Avocado is mainly used fresh, in sandwich filling or salads when they are in the ripe state. It is additionally used in ice creams and milkshakes and the pulp may further be preserved by freezing. Avocados can be grown on a wide range of soils, but they are extremely sensitive to poor drainage and cannot withstand waterlogging. Nutritionally avocados are a great source of vitamins C, E, K and B-6 as well as riboflavin, niacin, folate, pantothenic acid, magnesium, and potassium (Gururaj et al., 2017). They also provide lutein, beta-carotene, and omega-3 fatty acids. Although most of the calories in an avocado come from fat, they are full of healthy, beneficial fats that help to keep one full and energetic.

Once consumed, they turn off the appetite which aids a person in losing weight, by slowing down the breakdown of carbohydrates in one's body and keeps the sugar levels in the blood at balance. Eating the healthy fats from the avocado fruit enhances a supports skin health, increases the absorption of fat-soluble vitamins, minerals, and other nutrients, and may even help boost the immune system in humans. It is healthy for the heart as it helps maintain healthy cholesterol levels. It also enhances vision and studies have shown that taken adequately protects against colon, stomach, pancreatic and cervical cancers. For pregnant mothers, it reduces the risks of miscarriage and neural tube defects. Besides all these benefits they also lower the risks of depression, improves digestion and is also a natural detoxifier where it regulates the immune system and inflammation.

The Jamun/Jewish Plum (*Syzygium cumini*) minor fruit is mainly cultivated in Southeast Asia but it has found its way too many markets including India, Thailand, Malaysia, Philippines, Indonesia, Sri Lanka, etc. This genus has about 15 species that are reported to produce edible fruits and show a high amount of genetic diversity. Jamun minor fruit has proved to be the most important among all species due to its high contribution towards nutritional values among mankind and animals. These fruits are a good source of iron, sugar, minerals and carbohydrates. Moreover, these fruits and seeds possess anti-diabetic compounds and are popularly used as medicine. *Syzygium jambos, Eugenia uniflora* and *Syzygium malaccense* fruits are also very popular and eaten as fresh. About 10 species and 15 varieties are maintained in the gene bank for further and future analysis.

Some of the medicinal values derived from these Jamun fruits included their anti-diabetic properties (Haque et al., 2018). They are also good for the digestive system because of their coolant features. Additionally, they are rich in antioxidants and have adequate amounts of iron and Vitamin-C. Besides, the Iron content in Jamun acts as a blood purifying agent besides, helping in the cure of jaundice and anemia in patients.

Figure 2.11: Jamun Plum (*Syzygium cumini*) source of iron, sugar, minerals and carbohydrates.

Source: *123RF*

The Ber *(Ziziphus mauritiana*) adds to another type of minor fruit that is mainly produced in South Asia. They are cultivated for a period of five months ranging from September to January. There are about three species of this genus that contain edible fruits. This ber fruit is commonly cultivated in India whereas Jujube is predominantly cultivated in China. Its plant is reported to be highly drought-resistant and as a gift to arid regions in most parts of the world. These fruits are eaten fresh while some are processed in order to prepare the traditionally known delicious murabba and candies. These fruits can also be used in the preparations of jam and wine. The varieties that have been maintained over the years for further analysis and genetic diversity studies are four. *Ziziphus mauritiana* fruit plants help in weight loss and also strengthens the immune system in human beings. Additionally, this genus once consumed soothes the nervous system with its sedative properties (Hoang, 2008). Past studies have shown that this ber fruit could help fight Alzheimer's.

Figure 2.12: Ber (*Ziziphus mauritiana*) be used in the preparations of jam and wine

Source: *Wikipedia*

The other type of minor fruit is the passion fruit (*Passiflora edulis Sims*), which is a native of Brazil belonging to the family Passifloraceae. It is produced mainly in tropical and sub-tropical parts of the world in South America to Australia, Asia and Africa. Passion fruit is grown on many soil types but light to heavy sandy loams of medium textures are most suitable. This fruit is perennial and climbing with a woody vine that produces ovoid, purple-colored fruits when ripe. The fruits are known to possess a unique flavor and aroma and are also known for their high nutritional and medicinal properties (Hossain et al., 2017). These fruits are processed to make fruit juice and concentrate for domestic and commercial purposes. The Fruits are rich in Vitamin A, Vitamin C and minerals such as sodium, magnesium, Sulphur and chlorides. Passion fruit is used medicinally for urinary infections and as well as a mild diuretic, digestive stimulant, and heart tonic and to treat asthma, whooping cough, bronchitis, coughs, gastric cancer, etc. The two types of the genus are the standard yellow (*Passiflora edulis* f. flavicarpaDeg.) and the purple (*Passiflora edulis* f. edulis). The yellow is more acidic and less starchy while the purple is less acidic and starchier.

Flowers

Fruit on the vine

Yellow passion fruits and its cross-section, showing seeds

Purple passion fruit and its cross-section showing seeds

Figure 2.13: Passion fruit (*Passiflora edulis*Sims) is used medicinally for urinary infections and as well as a mild diuretic, digestive stimulant, and heart tonic and to treat asthma, whooping cough, bronchitis, coughs, gastric cancer ,etc.

Source: *Frontiers*

Drum Stick (*Moringa oleifera*) is the other type of minor fruit that is perennial in nature and is said to be a multipurpose tree. The Drum stick is popularly grown in home gardens in areas such as Sri Lanka, India, Indonesia, Thailand, Vietnam, etc. All parts of this tress possess at least one or two nutritional elements. To start with the leaves are a rich source of iron, fiber and Vitamin-A. They are additionally known as food for anemic patients, women with children and the old. Their pods are produced four times a year and used as vegetables (Hilson, 2016). These pods are a rich source of vitamin-A & C, iron, calcium and potassium. Besides, the drum stick Seeds are used for the extraction of high-value oil. All the genetic variability has however been collected widely for further studies. The drum Stick mineral contents are abundant. They are said to have 7 times the vitamin C in oranges, 4 times the calcium in milk, 4 times the vitamin A in carrots, double the proteins in milk and thrice the potassium in bananas.

Figure 2.14: Drum Stick (*Moringa oleifera*) to have 7 times the vitamins C in oranges, 4 times the calcium in milk, 4 times the vitamin A in carrots, double the proteins in milk and thrice the potassium in bananas.

Source: *news prestiti*

The tamarind (*Tamarindus indica*) is also an indigenous minor tropical fruit found mostly in tropical Africa and South India. The species of this genus are highly drought-resistant and are majorly grown in marginal and wastelands. When they are unripe they are used as an ingredient in vegetable dishes while the ripe fruits are a rich source of Vitamin-C and tartaric acid. Tamarind plant pulp is an important ingredient in several food preparations and the Seeds are used for the extraction of starch, protein and oil there are four genotypes of Tamarind.

Pomelo fruit has the largest fruit among citrus fruits in the minor fruits. It belongs to the family Rutaceae. Its botanical name is *Citrus grandis*. Locally it is known as Chakotha, Chakotaraetc. In India this is grown on a limited scale, in few parts of the countries especially parts of Karnataka, near the Kanya Kumari region of Tamil Nadu and lower hills of Uttarakhand land Himachal and parts of Bihar. This fruit has good potential for sub-tropical climate but lack of good varieties and lack of suitable varieties, poor marketing; it failed to compete with other crops. But due to increased health awareness and attraction of people towards less utilized fruits, the popularity of this crop is projected to increase. Pomelo fruit tastes similar to grapefruit however it's sweeter. Past studies have shown this plant to contain s several vitamins, minerals, and antioxidants that make it a healthy addition to the human's diet (Islam et al., 2019). This fruit is an excellent source of vitamin C is a powerful immune-boosting antioxidant that helps prevent cellular damage from harmful compounds called free radicals. It is also rich in potassium, which helps regulate fluid balance and blood pressure. It is also said to be very rich in fiber, which helps add bulk to stool and prevents incidences of constipation in humans. Nutritionally the fiber also acts as a source of food for the healthy bacteria living in the gut part of the human body. The fruit fiber has been associated with improved bone density, long-term weight maintenance, improved gut and brain health, and a decreased risk of some chronic diseases. It is also additionally rich in antioxidants, boosts heart health, fights cancer cells and also possesses anti-aging properties.

Figure 2.15: Pumelo fruit rich in potassium, which helps regulate fluid balance and blood pressure.

Source: *YouTube*

Macadamia (*Macadamia integrifolia*) is another type of minor fruits. It is a delicious and nutritious nut indigenous to the eastern Australian part of the world. The edible part is the white cream-colored kernel or embryo which is enclosed in one piece of very tough spherical brown seed coat or shell which is surrounded by a green, fibrous pericarp or husk. Its tastes resemble the cashew nut taste but, it is oilier and globular in shape. The kernels contain relatively large amounts of protein and oil. Macadamia nuts are considered a good source of calcium, phosphorus and iron. They are linked with other benefits, including improved digestion, heart health, weight management and blood sugar control. They are calorie-rich nuts that are high in healthy fats, vitamins and minerals. The macadamia nuts are loaded with antioxidants that fight inflammation and help lower cholesterol levels in human bodies (Jamnadass et al., 2011). Consumption of macadamia may lower the risk of heart diseases such as stroke and fatal heart attacks. This fruit may reduce the risk of metabolic syndrome, improve gut health, helps in reducing weight and have other potential benefits including anti-cancer properties.

Figure 2.16: Macadamia (*Macadamia integrifolia*) contains benefits which include improving digestion, heart health, weight management and blood sugar control.

Source: *Daleys Fruit Tree Nursery*

Moreover, there is the *Garcinia gunmigatta* (Cambogia). It is also known as kodumpuli or Malabar tamarind. It is an indigenous type of species fruit that is mainly used as species in curries and its juice is used for reducing fat and also prevents blood clotting by reducing triglycerides. Additionally, in South Tamilnadu it is used as fish preservative. When ripe the fruits are processed and used in the preparation of fish and prawn to impact the flavor and the taste while it improves the shelf life. It has immense value in drug production for reducing obesity and rinsing the mouth as it removes bad odor in humans (Kotwal et al., 2011). This genus seeds contain edible fats and the tree additionally yields translucent yellow resin, which is used as a varnish because it has purgative properties. The bark and fruit rind decoction are used against rheumatism and bowel complaints and are good vermifuge. The dried fruits and leaves are used for curing diarrhea, dysentery and dyspepsia. They too contain anti oxidizes that are very essential to any human being as they prevent sicknesses.

Lastly is the dragon fruit that is dark red.it is quite important to the human in treating skin problems such as acne and aging. It also helps in regulating cholesterol levels, high blood pressure and blood sugar levels.

2.5 MINOR FRUITS FOR SOURCE OF SUSTAINABLE INCOME

In rural areas where the majority of the human population is resource-poor, minor fruits can act as a source of sustainable income. In general, some of the plant parts including, the roots, leaves, gum, fruits among other parts tend to be of high economic value in the country. The under-recognized members of the public including women and children and those people that are found in the weaker sections of the society. Some of their collection of fruits are retained in the community for home consumption. These fruits are either eaten in dried form or raw and only when the fruits are in excess, they sell them in the market (Koohafkan & Altierim, 2011). The market of the minor fruits tends to be influenced by several factors such as value addition, processability, export, foreign demand, domestic consumption which are always changing based on the knowledge that farmers know about the help promoting qualities and the nutritional values of such fruits. In most cases, minor fruits tend to be eaten as is but not processed.

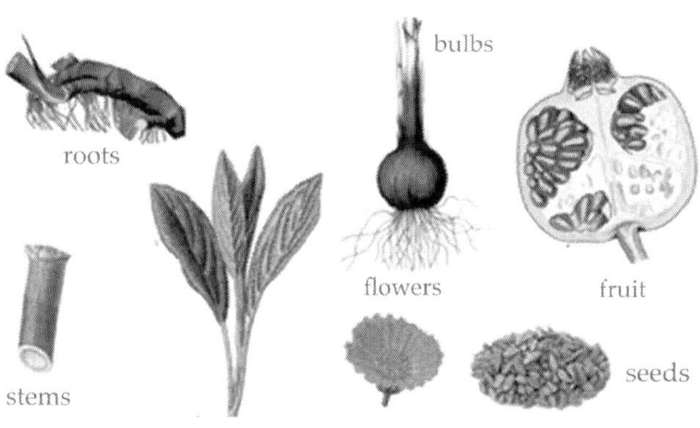

Figure 2.17: The majority of plant parts have high economic value in most countries.

Source: *Plants for a Future*

Chapter 3

VALUE ADDITION IN MINOR FRUIT CROPS

CONTENTS

3.1 INTRODUCTION

Minor fruit crops can be characterized as those farm products which however can still be consumed by individuals yet are moderately less tasteful compared to other standard farm produce, which has lesser interest on the lookout, are developed partly just, and are not normally developed in coordinated manors with the utilization of data sources.

Different terms that are utilized for these products are lesser-known farm products, underutilized organic products, less engaging farm products, under-misused products, likely natural products, stray natural products, wild food products, and so on. Minor organic product species go about as life uphold species in outrageous natural conditions and compromised environments and have resistance to make due under brutal environmental conditions.

IIf minor fruit crops are appropriately used, they may assist with contributing to food security, nourishment, wellbeing, pay age, and natural administrations (Kalinganire et al., 2007). The minor harvests species are regularly native old yield species which are as yet utilized to a certain extent inside the neighborhood, public, or even global networks, however, can offer more towards the accessibility of food sources than they as of now do.

The majority of nearby individuals know about the restorative properties of privately developed minor organic products. Besides, these harvests are assuming an imperative part in the sustenance and vocation of country and ancestral masses for work and pay age. The vast majority of the minor organic products are exceptionally valued by the shoppers and have interesting potential for esteem adding through preparing and item improvement.

All in all, crops like natural products are a rich source of nutrients and minerals. Backwoods products of the soil are a portion of the underutilized crops regardless of being utilized for human utilization, and monetary purposes. Utilization of different sort of occasional natural products is useful in keeping people sound and good by satisfying their different necessities.

In any case, various underutilized organic products have extraordinary restorative qualities. An examination of these natural product plants concerning their utilization in the infirmity of humankind will demonstrate productivity in the occasions to come.

Agricultural nations are being urged to expand their food sends out by growing new items and increasing the value of items already in existence. Worth expansion and enhancing food sends out depends not just on evolving creation and preparing frameworks yet besides on connecting into suitable showcasing networks?

For example, in territories nearby the backwoods, a large number of individuals accumulate minor wood farm products. A significant piece of these produces are for family unit utilization and pay depends on woods gathering, anyway rural exercises giving strengthening pay (Krishna et al., 2019). Woods nourishments are healthfully significant and are customarily utilized as enhancements to the staple eating routine.

Green vegetables and creatures in the wild add variety, flavor, nutrients, and minerals to typically grain overwhelmed consumes fewer calories. Privately harvested underutilized little millets, and vegetables also, native organic products give food just as money pay for native individuals and are critical in guaranteeing worldwide food security. Others likewise assume a critical part in keeping up the profitability and strength of conventional agricultural ecosystems.

These mostly minor horticulture crops also, a wide assortment of native leafy foods add to food security and assume indispensable parts in the sustenance that advance the eating routine of the provincial people.

3.2 VALUE ADDITION FOR AGRICULTURE

Value added horticulture alludes majorly by and large to assembling measures that expand the estimation of essential agrarian products. Worth added agribusiness may likewise allude to expanding the financial estimation of an item through specific creation measures, e.g., natural products, or through provincially marked items expanding customer allure and ability to pay a premium over comparative however items that are not differentiated.

Its portrayal can likewise be as the cycle that changes the crude horticultural item into something new through bundling, preparing, cooling, drying, extricating, and different cycles that change an item from its unique crude structure. Because of this change, the client base of an item and income hotspots for the maker is extended.

For this technique, cultivating is not, at this point limited to the development of an immense parcel of land or the consideration for countless creatures to be productive. In this case, even the individuals who not exactly a section of land could accomplish practical cultivating basically by expanding an agrarian item's potential with the goal that its saleability is upgraded (Marín et al., 2008).

If a maker cultivates strawberries, for instance, he ought to not just sell the harvest as new berries since he could likewise benefit from its different bits by creating different items, for example, strawberry jam and syrup.

This idea has acquired money in the little homestead strategy banter, in light of the worry that the ranch estimation of the shopper food dollar keeps on diminishing. Value added agribusiness may be for ranchers to catch a bigger portion of the buyer's money set aside for food.

Models incorporate direct promoting; rancher responsibility for offices; and creating ranch items with a higher inborn value, for example, character saved grains, natural produce, natural meat, unfenced chickens; and so on, given purchasers are eager to follow through on a greater expense than for more customary mass wares.

For better yields, the key is the appeal in item specialties and this is shown in the ubiquity of items, for example, garlic meshes, a sacked plate of mixed greens blend, craftsman bread, lavender cleansers, and hotdogs, among others.

Value added agribusiness is viewed by a section of individuals as a huge provincial advancement procedure. This is viewed as a path for makers not exclusively in the United States, to build their pay yet besides rival ware advertisers and upstream processors (Pingali, 2007).

Limited scope, natural food handling, non-customary yield creation, agricultural travel industry, and biological energy advancement are instances of different value added projects that have likewise made new openings in some provincial spots.

The expanding market division supports this positive condition which makes an uptick in the interest for high-caliber and separated horticultural items. As per an investigation, esteem added horticulture permits ranchers to catch as much as 80% of the estimation of a large portion of their rural produce.

Advantages of value-added products

- More volume can be dealt with per unit territory.
- Try not to confer shading to the final result (particularly on account of flavor oil).
- Not influenced by bacterial defilement.
- Simpler to store.

- The time span of usability is longer when contrasted with straight zest.
- It builds the turn of events and development of
- The united business and along these lines gives extra work opportunity.
- An item can be created.
- Specially designed blinds are likewise offered to suit the particular prerequisite of the client.
- Sterile and liberated from microorganisms.
- Contain normal cell nutrients and are free from compounds.

3.3 THE NEED FOR PROCESSING AND VALUE ADDITION

Notwithstanding the immense hereditary variety of natural products, just significant organic products like mango, banana, and citrus have been acquired in the profitability and adequacy of the individuals. A large number of the native products of the soil are still underexploited because of the absence of familiarity with their latent capacity, market interest, and low and unpredictable bearing as a rule.

The modernization and enhancement of present farming and ranger service-related exercises upheld by proficient handling of the items with the end goal of significant value expansion are required to build food creation, make business, and produce pay thus by increasing the value of the food items.

After collecting, it will limit the misfortunes during safeguarding and handling of these produces may bring more advantages, bringing about additional pay to the ancestral farmers or processors, in business advancement and this improves the nature of products to the purchaser (Sarker et al., 2017). A portion of these yields are likewise amiable to be developed alongside contending money yields and in this way supplement the local food creation.

The special power of specific underutilized disregarded yields in their rich and ideal nourishing arrangement, nutraceutical value, and item improvement offers extraordinary freedoms for money to the farmers, specifically the homestead ladies. The general effect of such intercessions on the gatherings of harvests to the economics of the networks saving and developing them, upgrading their food and dietary security, and fortifying their conventional food culture.

On account of the underutilized and dismissed plant species, markets and market-related issues, for instance, absence of market access, powerless market interest, helpless market structure, market motivators, and inappropriate market channels have been distinguished as examples of the numerous worries that small-scale farmers face.

3.4 PREPARATION OF VALUE-ADDED PRODUCTS THROUGH PRESERVATION

Conservation is a cycle of placing food materials in an adjusted condition for quite a while that does not hinder their quality to the most extreme degree, with the goals to safeguard leafy foods at the phase of greatest satisfactoriness, taste, shading, flavor, quality, and nutrient value.

It also looks out for wastage of nearby or occasional excess; to make the item accessible for a more drawn out period even in where it isn't delivered; to protect food materials during travel from maker to shopper, and to encourage treatment of food substances (Srivastava & Sarkar, 2017). This is fundamentally done through different strategies for bundling.

Agriculture is worried about short-lived yields, for example, foods grown from the ground. Except if the safeguarding business creates with the advancement of agriculture, it will not be friendly to the economy. Conservation deals with excess produce and in this way checks wastage from decaying: that is, the more we save, the more we can consume later on. Conservation causes the farmers to acquire a better return by checking wastage during a market overabundance.

The items are kept in appropriate conditions, which is beyond the realm of imagination under common or cold stockpiling conditions. Henceforth, protection is a reasonable substitute for capacity. It permits leafy foods to be accessible during slow times of the year and in areas where these are not developed.

Likewise, conservation has the extra advantage of food sources being more attractive. A few auxiliary ventures, for example, the creations of jars, bottles, covers, cardboard, and so forth might be set up, which will produce business openings (Sahoo et al., 2019).

3.4.1 Scope of Fruit Preservation

Leafy foods and fruits are significant enhancements to the human eating regimen, as they give minerals and nutrients fundamental to keeping up wellbeing and shielding us from various infections and issues. As indicated by human dieticians, a grown-up individual working decently requires 85 g of natural products also, 300 g of vegetables consistently, notwithstanding grains, fish, meat, milk, and so on.

A nation like, India, with its wide scope of soil and climatic conditions, is essential for the development of leafy foods and is the second-biggest maker of products of the soil, creating about 81.2 million tons of foods grown from the ground million tons of vegetables, of which around 25 % to 30 % of the absolute produce is squandered due to decay (Sayer et al., 2012).

Most products of the soil are naturally occasional and transitory. At the peak harvest time, there might be a market excess, but since of inadequate transport and lack of access to packing materials, the excess can't be taken rapidly to the business sectors in metropolitan zones.

Additionally, the excess regularly can't be put away available to be purchased in the slow time of year in light of deficient nearby cool storage spaces; along these lines, the cultivators don't get a decent cost for their produce because of the excess, and some of it is ruined, bringing about complete misfortune.

Safeguarding of products of the soil can assist with taking care of these issues. Little, inadequately molded, overripe, and tainted products of the soil that are unsatisfactory on the lookout and get a lower cost can be used effectively in the safeguarding business. With expanded urbanization, ascend in working-class buying power, alterations in food propensities, and abatement in the act of creating protected items in individual homes, there is expanding interest for industry-made items in the homegrown market.

Additionally, a portion of these safeguarded items, for example, canned mangoes, natural product juices, salted cashews, dried out nourishments, and frozen natural products are acquiring fame in the unfamiliar market and are acceptable to unfamiliar trade workers (Shackleton et al., 2007). Notwithstanding every one of these reasons, just 2.2 % of the absolute produce is prepared in India when contrasted with 40 to 83 % in created nations.

3.4.2 Limitations of the Fruit Preservation Industry

Absence of coordination among producers and preparing units. It is important to provide products of the soil ceaselessly for the preparation of enterprises to adequately run. An agreement among producers and the preparing units guarantees proceeded with the accessibility of good quality crude materials to the business.

Absence of talented labor. Albeit most created nations have a worthwhile situation in possessing a huge supply of labor, gifted labor is not inadequate stock. Laborers ought to be appropriately prepared.

Absence of mindfulness. Since the vast majority of the accessible information concerning conservation is dispersed in logical articles, reports from departments, and other profoundly specialized distributions, individuals when all is said in done don't know about the cutting edge strategies and methods of protection. To defeat this issue, this data must be scattered on a nationwide scale with the goal that the full favorable position of it tends to be taken by each one of those inspired by the business (Singh et al., 2014).

Absence of advertising resources. Even though there is an interest in preserved items, there are not promptly accessible in humble communities since retailers are reluctant to stock such things. The foundation of a cultivators' agreeable society would help in the promoting of such items.

Trouble in the accessibility of compartments. Jugs and jars are the two significant sorts of compartments needed by the food handling industry. The activity has just been taken up by various plants for assembling containers of required detail, yet there is incredible trouble in the accessibility of jars because there are not many plants for their assembling (Tiwari, et al., 2018). It is necessary to set up more manufacturing plants to fulfill the need for jars.

3.4.3 Value addition in Underutilized Fruits

The differing climate states of this nation give a reasonable climate for developing an assortment of natural products. Food scientists suggest that organic products should shape a significant constituent of the everyday diet. Various organic products' tones, differed smell and taste make them acceptable and refresh over brain and body.

The significance of natural products depends on their nutritive value, yet additionally in the way that it tends to be used crude and whenever of the day. Utilization of organic product relies upon its general creation and utilization by people and food ventures. Based on these models, it created the impression that even the most regular natural products are under-used. Tropical organic products which are presently not well utilized, have a significant task to carry out in fulfilling the requests for nutrient-rich, carefully enhanced, and alluring regular food sources of high restorative value (Tiwari, et al., 2018). Less well known not well utilized natural products have been generally consumed as a staple food and for therapeutic purposes.

Organic products like bael, aonla, phalsa, karonda, custardapple, ber, lasoora, jackfruit, and so forth are of incredible significance because of their restorative properties. Only a little amount of these natural products is utilized in some drug enterprises. These organic products are the principle source of dietary fiber for the poor furthermore, hence assume a significant part in surviving the issue of hunger.

Handling is the most ideal method of using overflow creation of organic products during occasional excesses (Hossain et al., 2017). Handling helps in changing over transitory organic products into sturdy structure natural products, which are extremely hard to eat insanely can be prepared into a scope of the exceptionally worthy natural product item. There is an incredible breadth of food handling and different items like jam, jam, safeguard, squash, treats, pickle, and dried items.

Importance of Minor Fruits in Food and Livelihood Security

Food security implies that all individuals, consistently, have physical, social, and monetary admittance to adequate, protected, and nutritious food that meets their dietary necessities and food inclinations for a functioning and solid life.

Work security then again is characterized as satisfactory and supportable admittance to pay and assets to address essential issues including sufficient admittance to food, consumable water, wellbeing offices, instructive freedoms, lodging, and time for local area cooperation and social incorporation.

Jobs are often comprised of the scope of on-ranch and off-ranch exercises that together give an assortment of obtainment procedures for food and money (Legwaila et al., 2011). Subsequently, every family unit can have a few potential sources of privilege, which establish its vocation. Given the expanding populace pressure, agricultural nations face genuine difficulties of food security, joblessness, and adverse climate.

Staple harvests face significant difficulties soon and an expansion away from over-reliance on staple yields will be significant as a component of the advancement towards the objective of accomplishing security of food creation.

Nowadays, shoppers are getting progressively aware of the wellbeing and nourishing parts of their food containers. The inclination is to keep away from synthetic compounds and engineered nourishments and inclination for sustenance through normal sources.

The less utilized natural products are the principle sources of business for poor people and assume a significant part in conquering the issue of ailing health. The minor organic product crops assume a significant part in guaranteeing the vocation security of rural networks by at least one of the accompanying means:

- Decrease the danger of over-reliance on exceptionally restricted quantities of significant staple food crops.
- Assist the poor as a means of resource acquisition and pay.
- Increment maintainability of horticulture through a decrease in information sources.
- Assisting in saving, and praising social and dietary variety. Utilize peripheral and badlands for farming purposes to match the rising food demand.

Consequently, minor organic product plants give immense freedom to job improvement and local area advancement. Natural products have an incredible financial potential and assume a huge part in rural territories by giving supplement beneficial consumer routines and creating side pay to the destitute individuals (Malik et al., 2013).

A huge number of individuals, generally in non-industrial nations, infer a significant piece of their means and pay from these wild plant items. Minor natural products give major foods to native individuals, fill in as reciprocal nourishment for non-native individuals, and offer an elective source of money pay.

It has been widely established that the wealth of such natural product plants in the wild territory can be monetarily developed easily. Healthfully rich natural product items as jam, jam, juice, and pickle in such cases can be fabricated to create pay for poor rural individuals and diminish their financials and job trouble. For such manageable cycles, appropriate preservation practices and strategies should be formed as well.

3.5 JACK FRUIT

Jackfruit has a place with the Moraceae family local to south India. Jackfruit is a tropical organic product perceived for its novel shape and size (Mahapatra & Panda, 2012). Jackfruit is an indispensable piece of basic Indian eating routine and is openly accessible in India and its connecting mainland, its restorative properties are additionally referenced in Ayurveda.

The needy individuals of jack organic product developing a territory used to eat this natural product rather than rice, for one of the everyday dinners. Consequently, a jack organic product is called 'helpless man's food'. Individuals burned-through it generally as a natural product when ready yet additionally as a vegetable when it has not yet reached the ripening stage.

A Jack natural product fundamentally adds to the sustenance of the individuals of this country as a source of nutrients, minerals, and calories. Both delicate and ready organic products just as the seeds are plentiful in minerals and nutrients. It is developed and sold in the market practically entire the year in the country. In any case, the natural product is transient and can't be put away for a long time due to its intrinsic both compositional and textural.

Figure 3.1. Jackfruit tree with fruits.

Source: https://en.wikipedia.org/wiki/Jackfruit

3.5.1 Value Added Products from Unripe Jack Fruit

Chips. Crude or unripe jackfruits are utilized for making ready chips. Cuts of reasonable sizes are chopped and whitened into high temp water at 95°C for 5 minutes. At that point, the cuts are dried at 60°C for 1 hour and 70°C for the next 6 hours (Mbuvi & Boon, 2009). At last, the cuts are seared at 160°C in palatable oil. The cuts are taken out from the skillet when they turn light yellow in shading and blended in with salt and pressed.

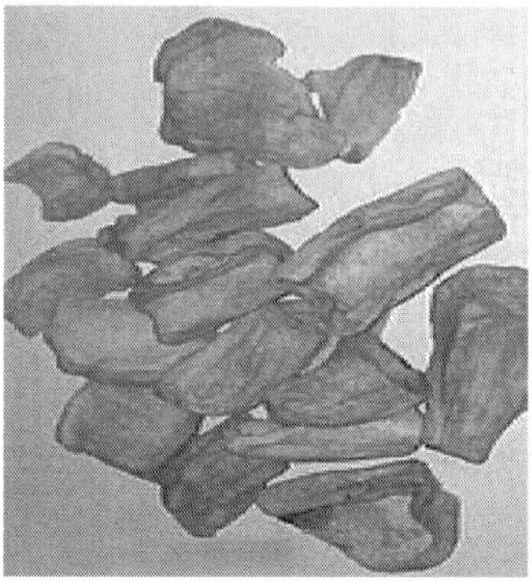

Figure 3.2. Jackfruit chips.

Source: *https://en.wikipedia.org/wiki/Jackfruit*

Pickle. Jackfruits in their unripe form are utilized for making pickles. Little pieces are produced using bulbs and seeds and they are blended in with the oil, salt, and flavors prior to pressing.

Brined product. Develop green jackfruits that are properly washed under sanitary conditions, stripped, and cut into little pieces. At that point they are kept in 8% salt, 1.25% acid, 0.1% KMS, and 91.65% water mix. At that point, the materials are filled with impermeable plastic holders and put away in a cool and dry spot.

Ready to cook jack fruit. After washing the jackfruit, it is stripped and cut into reasonably estimated pieces. These pieces can be saved and bundled with insignificant handling to plan prepared to cook jackfruit items.

Dehydrated jack fruit. The unripe develop bulbs can be whitened and get dried out for additional utilization consistently.

3.5.2 Value Added from Ripe Fruit Jack

The recuperation of bulbs or eatable segment (pulp) from jackfruit shifts from 20% to 25%. The bulbs from the organic products are taken out physically in the wake of cutting the natural product into a few pieces. As the natural product contains tacky latex, a little amount of vegetable oil is applied and the seeds are taken out from bulbs.

Jam. It is set up from the pulp of ready natural products with added substances. Bulbs from a completely ready jackfruit are mixed and bubbled for 5-7 minutes to separate the juice. At that point, 700 g sugar and 10 g gelatin are added to 1 kg jackfruit pulp and cooked until the TSS scopes to 64° Brix, at that point citrus extract (0.25%) is added (Mishra et al., 2018). Endpoint is resolved through a drop test and the jam is transferred while hot in a sanitized container and put away at room temperature.

Jackfruit rind jelly. Completely developed ready jackfruit are collected and washed with clean water. The skin is isolated and cut into little pieces. At that point, 1.5 liters of water and 2.0 g citrus extract are added for every kg of skin and the substances are bubbled for 35 minutes and the juice is removed from the skin. 700 g sugar and 200 mg citrus extract are added to this juice, and the juice is then cooked and this is followed by the option of citrus extract. The jam is then filled in a disinfected bottle and put away at room temperature.

Jackfruit jelly. Completely ready jackfruit is washed and bulbs are taken subsequent to eliminate the seeds. These bulbs are mixed with 10-15% sugar and bubbled for 5-7 minutes for separating juice. After this KMS at 0.1g/kg is added and afterward further bubbled for three to five minutes. The blend is concentrated with a steam jacketed container and spread in a tempered steel plate. The plate is placed in a cupboard dryer and dried for 20 hours at 60°C until the dampness content drops (Prabha, & Kumar, 2015). Subsequent to cooling, it is cut into bits of alluring sizes and put away for packing.

Nectar. The bulbs go through the pulp machinery and pulp is made by blending around 10% boiling water. Nectar is set up from this pulp.

Canning. Canning is done for the unripe bulbs with a little amount of citrus extract while the ready bulbs can be safeguarded in syrup made of sugar or as an improved pulp for a one-year period. The mash can be utilized for planning other value added items such as squash, prepared to serve drink, chutney, toffee and so on and for flavored frozen yogurts, custard, refreshments and pastry shop items.

Figure 3.3. Jack fruit added as an ice cream topping.

Source: https://en.wikipedia.org/wiki/Jackfruit

Readily consumable products. Ready jackfruit bulbs can likewise be safeguarded with insignificant handling into a "prepared to eat" comfort food item. In any case, this item has a restricted time span of usability and must be put away and shipped under refrigerated conditions.

3.6 TAMARIND

Tamarind pulp powder. Tamarind, half-developed is reasonable for the planning of tamarind powder. It is by and large set up by drying the organic product pulp in the dryer particularly in a bureau dryer at 50–60 °C up to the time it arrives at a humidity level of 5 %. In the wake of drying, the powder is set up from the pulp. The shade of tamarind is because of the presence of anthocyanin. It is profoundly solvent in water and used to confer the raspberry red tone in diverse food items. Tamarind powder is utilized as a red colorant in rasam, sambar, jam, and so forth.

Tamarind concentrate. For the arrangement of tamarind concentrate, the pulp is warmed with bubbling water, and solvent solids are separated to the degree of 20 % utilizing countercurrent standard. The product is then sieved and gathered under a vacuum in a constrained flow evaporator and directly filled in jars or containers (Pandit & Acharya, 2008). The strategy is created by CFTRI, Mysore. Concentrate from the tamarind sets like jam in the wake of cooling. The solvent strong substance is around 68 %. It has a greater shelf life than tamarind pulp. Newly gathered natural product pulp gives great quality items. The concentrate is made up of tartaric acid (13 %), simple sugar (50 %), gelatin (2 %), protein (3 %), cellulosic material (2 %) and water (30 %).

Figure 3.4. Tamarind paste.

Source: *https://en.wikipedia.org/wiki/Tamarind*

3.7 KARONDA FRUIT

Karonda has an extraordinary probability for esteem option and a few value added items can be set up from unripe just as ready organic products. A portion of the value added handled items that can be set up from karonda are examined beneath with their technique for arrangements.

Karonda candy. For making sweets, the cut or entire unripe product of karonda is whitened in 500 ppm potassium metabisulphite with heated water treatment. At that point, these readied cuts or natural product are saturated with syrup of 60, 65, 70°Brix with the expansion of citrus extract which contains genuine sweetener and jaggery at various levels. At first set at 60°Brix, the sugar and jaggery arrangement. Following 24 hours of soaking in every treatment, the syrups are depleted and their concentration were expanded by °Brix each time until the mix came to up to 70°Brix. The cuts or entire natural product according to treatment processed in every treatment was removed from the syrup and washed quickly with the faucet water and dried in shade for 24 hours. Subsequent to drying, the candy was stuffed in 250 gauge synthetic sacks and put away under optimal storage.

Karonda jam. Diverse hued products of karonda from purple to dark red are accessible in countries like India which are utilized to make jam. The ready organic product is brimming with acids and miniature and large scale supplements which join well with sugars and used to set up an assortment of jams. The natural products have a critical measure of gelatin consequently it is reasonable for making commercial jam. New and intact karonda organic products are washed appropriately and cut into equal parts. The seeds are taken out and the organic products are put in a weighty base container filled with water. The organic products are bubbled in water up to the point they turn delicate (Saha & Sundriyal, 2012). At that point, sugar is added and mixing is proceeded up to the end point decided via a drop test and by sheet test. For the readiness of smooth sticks, the delicate organic product goes through a sifter so a smooth pulp is acquired and afterward sugar is added. Once cooled, pack it in a glass container.

3.8 CUSTARD APPLE

Custard apple is perhaps the most tasty organic plant of tropical and sub-tropical regions and has a place with the Annonaceae family. In certain nations, natural products are eaten for the most part by individuals from the low and medium class. The gathering of custard apples is done in portions. However, the best harvesting stage is the point at which the firm organic product starts to create shading. It is for the most part picked when it gets velvety yellow between the portions and starts to break marginally. The natural product tends to blast open whenever kept on the tree for quite a while. Custard apple is exceptionally transient and can't be put away for a lengthy period. The ready organic products are high in sugar.

Custard apples are sweet and flavorful thus is utilized for table reasons. When completely ready, it is delicate to contact and handily harvested. The tissue might be scooped from the skin and eaten with no guarantees or presented with light cream and is sugary.

Shakes or smoothies can be processed from custard apples or even into normally frozen yogurts, Often it is squeezed through a strainer and added to drain shakes, custards or frozen yogurts. A heavenly sauce for cake and puddings can be made by mixing the cultivated tissue with crushed banana and a little cream.

3.9 POMEGRANATE

Pomegranate demonstrates the incredible breadth for the handling of items with added value having expanded time span of usability. The new area of pomegranate preparing permits the utilization of the natural produce with inferior quality organic products that can't be marketed, for the readiness of the new items. The movement of innovative work on pomegranate has focused on the utilization of new refrigerated advances to broaden the business timeframe of realistic usability of pomegranate.

The pomegranate is prepared into items like negligibly prepared new arils, juice, squash, refreshment, molasses, juice concentrates, frozen seeds, jam, jam, jellies, grenadine, wine, seeds in syrup, pomegranate spirits, pomegranate powder, pomegranate skin powder, anardana, ice cream parlor, and pomegranate seed oil (Schumacher, 2011).

Ready to eat pomegranate products. Recently, negligibly handled "prepared to-eat" pomegranate arils have gotten famous because of their comfort, high esteem, one-of-a-kind tactile attributes and medical advantages. Negligible handling of pomegranate comprises of washing with the purifying specialists to lessen the underlying microbial burden, pH adjustments, utilization of cancer prevention agents, changed environment packing and temperature control. The cultivar 'Mollar de Eche' produces the best outcomes. Advances, for example, latest packing and the utilization of common or nondestructive items as additives, (for example, nectar, aloe vera gel and chitosan and UV-C radiation) ought to be done in blend with MAP.

Frozen arils. after setting up the arils same as that of negligibly prepared arils, the arils were pressed in the polyethylene packs with syrup of 15°Brix. A chest cooler is used to freeze the arils. During freezing, the juice container with arils enters the syrup. The arils ought to be eaten frozen to stay away from an unreasonable loss of turgence.

Jams and preserves. An item referred to as anar rub with genuinely great maintenance of quality can be made by concentrating pomegranate juice and warming the combination on a moderate fire for a significant stretch. The wrapped-up item is thick and contains 70-75% TSS. It can be put away for one year and used as a jam and the best protection temperature revealed is 5°C.

Jellies. An alluring jam can be set up from pomegranate juice. While making jams, around half of the anthocyanins present in pomegranate juice are lost. On examination, it was discovered that the fermentation of juice delivered an essential enhancement in the shade of jam.

Anardana. Dried arils from acrid wild varities are called Anardana. It is utilized as acidulent instead of tamarind of dried green mango in North India in Indian styled curries, chutney and other meal arrangements. It is likewise utilized in the planning of stomach-related confections and by traditional arrangement of Ayurvedic and Unani medication) (Shajib et al., 2013). The improved preparing procedure comprises of pre-cleaning, motorized extraction of arils, sunlight based drying and bundling. After treating with sodium benzoate for 10 minutes, arils are dried out in a drier at 45°C for 48 hrs to 10-12% water content. It has the alluring earthy colored tone and can be put away for quite a while in glass containers.

Pomegranate molasses. It is a conventional Middle Eastern fixing produced using cooked-down pomegranate juice. Thick and sweet in surface, pomegranate molasses gives a tarty flavor and is dim in shading. Its pleasantness arises from the convergence of the organic product's regular sugars. To make ½ cup of molasses, 4 cups of juice is warmed in around five minutes, permitting it to thicken be that as it may, not overly cooked. The item can be put away in an impenetrable compartment under refrigerated conditions for a very long time. It is commonly used to season chutneys, curries and serving of mixed greens dressings to coat or soften meat items.

Pomegranate powder. Pomegranate arils that have been dried and processed are used to set up Low dampness pomegranate powder. The dynamic fixing considered in pomegranate powder is ellagic acid. The freeze-dried pomegranate powder rich in ellagic acid have numerous applications and might be typified for ordinary portions.

Pomegranate Wine. Wine Pomegranate juice can be used to make great quality wine. Sugar is added to change Brix to 22-23°. The sanitized juice is aged with starter wine yeast. The aging is permitted to proceed until wantedthe needed degree of liquor is acquired (Srivastava et al., 2017). The wine is clarified by bentonite treatment or by centrifugation. The wine matured similarly as red grape wine.

3.10 FIG

In antiquated civilization, the fig was a significant food crop. It is a profoundly nutritious organic product comprising of 84% pulp and 16% skin. In addition, the organic product likewise contains protein, calcium, iron, nutrient A and thiamine at different levels. The natural product is esteemed for its diuretic property. Therapeutically, it is applied for boils and other skin conditions. The latex is utilized to coagulate milk and leaves are utilized therapeutically for their diuretic, demulcent, emollient and anthelmintic properties. New figs are nutritious and utilized as a pastry or for making jam, jam, pudding, cakes, dried, sugar-coated or then again canned.

3.11 AONLA

The natural fruit is the most extravagant source of nutrient C and is a diuretic and purgative that helps in relieving sleep deprivation, scurvy, discharge, leucorrhoea, premature aging and utilized as a cooling agent to decrease the impacts of sunstrokes.

Aonla is additionally utilized in numerous hair products; it improves the development and hair pigmentation. In Ayurvedic applications, aonla powder and oil are generally utilized for the treatment of the scalp and improves composition, eliminates wrinkles and sunburns.

The supply of Aonla relies upon development at harvest. The natural product saves well in cool chamber for 17-18 days contrasted with 8-9 days at storing temperatures. In Ayurvedic planning like 'Chyavanprash' and triphala, Aonla is one of the primary materials. Organic product items like pickle, protect, sweets, jam, syrup and dried pieces are produced using Aonla.

3.12 JELLIES WITH GUAVA

Calyces of roselle can be utilized as a source of color for guava jam since there is an issue of searing in guava jam during capacity. Sensory assessment of jams arranged from various extents of guava-roselle uncovered the inclination of testers for mixing relatively to get jam with appealing tone (Tao et al., 2009). Roselle calyces can be utilized with guava for jam making without critical transformation in guava flavor.

3.13 PHALSA

Given that it is exceptionally transient; the natural product should be used within 24 hours subsequent to picking. The ubiquity of phalsa natural product is because of its alluring shading going from red to dim purple and its satisfying taste. It is famous and the extricated juice gives a profound ruby red to dim purple tone. It is evaluated extremely high in the native medication. The juice is very reviving and is considered to have a cooling impact particularly in blistering summer. Warming the squashed phalsa organic product to 50oC gives the most noteworthy recuperation of the juice with a fitting amount of anthocyanins and other solvent and insoluble materials.

3.14 BAEL

The bael organic produce is known for its therapeutic properties. The natural fruit is perhaps the most nutritious organic produce contains 1.19 mg of riboflavin/100 g consumable part. It could be noticed that no other fruit has this much concentration of riboflavin.

Bael fruits has been utilized from days of yore for handling in the development of a green structure to sour jam.The extraction of ready bael natural product pulp is determined by expansion of water equivalent in weight to the pulp, changing the pH to 4.3 with citrus extract and warming at 80oC briefly, before going through the pulper.

Expansion of water weakens the adhesive and the utilization of heat just inactivates the compounds yet in addition helps in dissolving the adhesive consistently all through the pulp (Tiwari et al., 2018). The fruit pulp subsequently got has nearly a similar thickness and shading as mango pulpjuice.

Ready Bael fruit juice, whenever extricated appropriately can be utilized for the arrangement of different natural product items viz., nectar squash, powder and so forth, which can be industrially used.

3.15 PASSION FRUIT

It is local to tropical America. The fruit is with exceptional flavor and smell for crisp eating. Energy naturalfruit has reasonable provitamin A, ascorbic acid, riboflavin and niacin and have a high mineral content.

The pulp got in the wake of scooping from the fruit at the point when sliced in equal parts are added to fruit plates of mixed greens, frozen yogurt or fruit juice. Other prepared items incorporate juices, and jam.

There exists minor fruit species that are wealthy in wholesome, therapeutic and preparing characteristics can assume an extremely critical part in the work security of the rural networks through upgraded family pay, business age and natural assurance. A large number of these minor nourishments have an incredible potential for business development.

It is necessary to expand the territory and creation of these lesse- known foods grown from the ground prepared items. Individuals from both developed and developing nations are getting progressively mindful about the likely significance of these minor natural fruit species in their food and job security (Viswanath et al., 2018).

Value expansion of these minor natural fruits for abusing their healthful and therapeutic properties is a drawn-out procedure and government and non-government associations should assume an essential part underway, showcasing, esteem expansion and promotion of these minor organic fruits.

Different partners such as the makers, neighborhood finders, vendos and laborers engaged with minor natural fruit farming, gathering, selling, shipping, and bundling may better their vocations from this area.

Additionally, improvement of current ease preparing innovations of significant worth option to be utilized for assembling expanded value-added items dependent on these minor natural fruits at little and house businesses level is likewise necessary.

Chapter 4

PROCESSING OF MINOR FRUITS

CONTENTS

4.1 INTRODUCTION

Minor fruit is a variation of the fruit cultivated on the way home, in the wasteland, and so on, scattered and unattended. Small fruits are generally regarded as human-consuming, but relatively less tasteful than other large fruits, with lower market demand and grow only to a limited degree, and are not typically cultivated by artificial agricultural income in organized crops.

They are usually long-lasting, well cultivated and have an intense capacity to exploit even in delicate soils and climates. Fatal injuries and epidemic attacks are extremely rare in all these cultures (Yadav et al., 2017). You replicate and develop naturally instead of dying out. In India, many unused fruits grow mostly without much thought at home, in the foothills and forest areas, and on roads and railroads.

The soil underneath one of these fruit trees is generally referred to as 'minor fruit.' They are responsible for the local climate; they play a significant role in herbal medicine and in the promotion of rural poverty and household food security. Although there are many under-utilized crops in different agro-ecological areas in India and other countries, nearly all of these small or under-utilized crops, some of which are eroded from the list of plants, there are no organized or approved crops.

Grocery shops are distributing fruit all over the country. In India, for example, a fruit has a greater part of the entire cultural scene. The production of a number of fruits in the area is estimated to be more than comparable to the previous year. Lost or lost due to lack of administration after harvesting. Complete losses from land to manufactured produce for consumption amount to as much as a few tens of millions of raw maximum value and then consumption.

There are many ways in which fruit and vegetables can be made more valuable and thereby decrease post-harvest losses and improve fruit consumption. This investment helps inspire customers around the world to take advantage of these pungent fruit cooking items over the coming year by stuffing small vegetables into a number of shelf-life extensions. Farmers may also help to counter waste by using their land, as they not only cultivate nutritious food, but also harvest the necessary fruit in the process. It gives you daily, warm food.

Diversification and production of marijuana, as well as the latest export of its essential ties between farmers and customers, was encouraged with the assistance of buyers and backers.

4.2 PROCESSING OF FRUITS

Although fruit should be considered as important dietary care for health benefits, it has the potential for foreign exchange revenues and added value. The fruit market has grown significantly in recent years and is now the cornerstone to the growth of our economy. Customers are involved in the well-being and health impact of the products in their food baskets. The goal of the initiative is to eliminate pesticides and synthetic chemicals from farm crops and to use natural nutrients.

The key suggestions to be adopted by the Government for the citizens in need and the resolution of the issue of hunger are in fruits such as kadu, rockets, Bael fruit, passion fruit, tamarind and apple (Yadav et al., 2018). While vitamins, minerals and dietary fibers are often called "healthy" and "nutritional," they are only at the lowest stage of the total nutritional diet.

The very serious loss of fruit after harvest is much more serious in developing countries than in developed countries. The overall losses between harvest and market-level are approximately 30 percent and 40 percent.

It's worth a thousand crores of cords. Processing of perishable goods and foodstuffs is required to ensure that farmers earn fair returns to boost their economic conditions. Processing is one way to deal with excess fruit during the off-season, to reduce surpluses and use them as a way to make the crop bigger. More equipment, both large and small, must be used during the market season to ensure that the plant continues to run longer and more presentable during shipment.

Food exports are extended through the supply chain, where the path or route or routes through which exports can be expanded or increased (Akinnifesi et al., 2002). Pomegranate is one of India's most useful fruit plants, due to its adaptive existence, high viability and commercial growth, and fruit is a good source of nutrients, primarily anthocyanins, which demonstrate high-quality chemical preventive action, including anti-mutagenicity, anti-hypertension, anti-oxidants and reduced liver function. Their products, in particular dry pears and processed fruit jellies, have distinct characteristics and palatability that make a food item safe for consumers.

4.3 EXAMPLES OF MINOR FRUITS AND HOW THEY ARE PROCESSED

4.3.1 Pomegranate

Figure 4.1: Pomegranates are known to produce good wine after processing.

Source: *https://timesofindia.indiatimes.com/life-style/food-news/unbelievable-benefits-of-pomegranate/articleshow/69940581.cms*

The pomegranate tree, also known as *Punica granatum*, is a deciduous shrub that grows between 5 and 10 meters (16 and 30 ft) in the Lythraceae family, the Punicoideae group. Granada, Turkish Young Side The pomegranate has many origins in northern India, particularly in the Iranian countryside, and has been cultivated in the Mediterranean since ancient times. It originated with the Spaniards at the end of the 16th century and was then brought to California by the Spaniards in 1769. The fruit season usually ends in the winter of the Northern Hemisphere from October to February. It's late spring in the southern hemisphere and summer in the northern hemisphere. As a result of frying, mixing juice, garnishing and alcoholic drinks such as cocktails and beer. The pomegranate is also found in intact sarcotes. It is now widely grown in Africa, the Indian subcontinent, the Middle East and the Caucasus, the driest zones, the Northern and Tropical Asian areas, the South East Asian zone and the drier regions. California,

Arizona, and the desert are both connected to the Central Valley. Smoking started to become popular in shops and markets in Europe and the West in the 20th and 21st centuries. The fruits have shown great potential to be used as a method of manufacturing value-added goods. Fruit with sunburn husks breaks and cracks and scald husks on the entire fruit will limit the marketability and perceived consistency of the fruit. There is a new trend in the use of pomegranate as a cheap alternative to indica berries in a supermarket selling fresh food items. The goal of the Granate Research and Development Operation was to use new cooling technology so that grenades can be kept at the correct level of freshness for longer (Ahmad et al., 2008). As modern grenades are not fully polished, they can be turned into things such as fresh arils, juice, squash, molasses and drinks that are minimally processed. We have eggs, fruit juice, nuts, almond butter, sauces, jams, syrups, anards, biscuits, cinnamon, onions, nutmeg, sumac, rosemary, fruit, elderflower, coffee, sherry, tamarind, etc.

Processing

Figure 4.2: There are a number of activities done during the processing of pomegranates.

Source: *https://www.youtube.com/watch?app=desktop&v=O2J3xxN4DW0*

Pomegranates have a particular fruit content of between 48% and 52% of the fruit. The most significant part of the fruit is the juice, which contains a high percentage of vitamin C. There is also a large seed in the fruit body. The seeds and arils are crushed and the juice is harvested and sold as fresh juice due to its excellent flavor, enticing scent, delicious taste, good nutritional and medicinal benefits. The processing of juice from pomegranate arils has become a valuable feature of one of the main ways of being useful. These liquors can be used to make many other things, such as syrup, nectar or gelatin. Wine made from the fruit's grape variety includes multiple coloring ingredients. The juice contains a rich supply of polyphenols. Antioxidants, astringents and tannins in pomegranate juice contain some of the polyphenol compounds in pomegranate, including anthocyanin, flavonoids and other polyphenols that give rise to bright color (Altendorf, 2018). Pomegranate juice has very high antioxidant effects, making it useful for health supplements and nutraceuticals.

Extraction of pomegranate juice

There are different methods used in extracting pomegranate juice. The biggest difficulty in the selection of flavorings is very tedious peeling, which can be cumbersome to the point of being unsanitary if the customer's hands are covered with polyphenols and oxidative enzymes. As a variety of fruits, this one is prepared to collect juices to soften the seed bags inside the fruit by rolling them to impress the hard surface. Mehak Shivpuri (Rahuri) developed a prototype computer that could strip the skin from the apple without damaging the fruit flesh inside. The gadget is made of titanium for CIPHET, Ludhiana, India, which fastens to yield fruit arils quickly. A direct method of juice extraction makes it possible to open the fruit, cut the pulp and either press it or place it in a screw or a basket press. The fruit is cut and fried. It is then pressed into a juice or pulp or pressed into a solid fruit block. Anyway, it's quartered and compressed. When the maturation cycle for the main pomegranate plant crop is complete, the remaining ripe fruits are packed (juiced) and sold.

It's great! It's better than that. The yield of juice per unit volume is roughly 42 per cent and the amount of juice obtained per unit volume is also 42 per cent. When it comes to weight gain, there's only 70% of the return. Ozkan has developed a process by which each pomegranate aril is extracted for pomegranate juice. As a final step in the process, the juice is passed through the gelatine and the filtering system. When the vendor forced the fruit out of the pulp, Saxena and his team got the juice out of the fruit by quartering it and melting it on the rack and the hydraulic press. The juice should be harvested at a pressure of less than 100 psi in order to keep the yield of rinder tannins from being too high (Banyal et al., 2019). As for Simon et al. Dr. Prescott, Dr. Spanos, and Dr. Ati became intrigued by the fact that the acidic, bitter and cloudy viscosity of fruit juices during concentration and storage is the phenol of fruit pomegranate juice. Adsule registered average recovery of hand-pressed pomegranate juice by weight of the patient. Production on both farms contributed to around 50% of the fruit. The yield of the seed was lower, but the yields were similar to those of the last years. The yield of grain was equal to the previous years and was similar to 75% of the global average. Chobe used a pulper to suck the juice out of the pomegranate pod. She then moved the juice into the juice extraction machine that she had rented to the grocery store. Since collecting the juice, she was able to achieve an average yield of 55%, making it easier to get enough juice for the wedding. Vardin and Fenercioglu are in the making.

The hand-operated packing press produced 30-40 percent of the pomegranate juice by pressing the whole pomegranate fruit in batches of 10 kg for 5 minutes. Pomegranates are studied in the same manner as a few additives that are pressed with powders, solids, frozen pomegranate pulp and the results were very strong. Quality pomegranate fruit is a fruit that the world has come to love, but the world is not getting any more. The flavor, nevertheless, induced inefficient astringency and bitterness due to the compression of phenolic compounds in the juice. Using the centrifuge method, Neifars et al. extracted the pomegranate juice at pH 4.1 and TSS 150Brix and distilled the juice at-20 °C using the electric juice of the Philips. When Alpher et al. tried to eat the entire fruit, they had to squeeze the two halves as closely as they could by hand to remove about sixty percent of the juice. Marti et al. argue that they have discovered a tool for extracting pomegranate juice. The ramping method was conducted using a potassium carbonate sedimentation stage (pH=3.8). The distillate was then removed by hand-squeezing or hand-holding apparatus, then by a commercial screw-style extractor or by hand-squeezing. The researcher estimated that the overall studies reported that the mean recovery was 81.0 percent in the overall studies. You may also put the arils in a cup of water for a few hours (Barua et al., 2019). It will no doubt be harmful. Red wine can be made with the aid of pomegranate juice. Apply the Brix to hit with a taste of 22-23°. Pasteurization is achieved by the fermentation of the juice.

Fermented beverage and fermentation chemicals. The fermentation process will continue until the end of the process is completed. You're going to get the number you want. Bentonite preparation tends to simplify the wine to such a degree that it immediately becomes pure. The wine was made a year in the same way that the red grapes are aged. As described earlier, it's dull and tedious as well as time-consuming.

The fruit develops while the fruit is prepared. The processing of pomegranate is tedious making the fruit of the table is unpopular. To address the many problems faced, several countries (including Spain, the United States and others) have taken steps to encourage the development of ready-to-eat pomegranate arils. Minimally made, ready-to-eat rings are very familiar to customers. The rings are rather simple and offer a high degree of appeal. The rings produce a unique sensory function as they create a low volume of noise and often provide a highly unhyped tactile sensation. In James Caleb, and so forth. In 2014, the lack of biological harmony, as it occurs in the natural world, was overcome by the activation of low-quality pesticides with surface skin defects, which often occur in the natural environment, including the use of low-quality fruits, which is also the case in the natural environment. Changes in food safety have been made in the management of post-harvest-specific problems, including pesticide activation on fruit with damaged skin, including splinters, sunburns and bacterial spots. It is a major challenge because pomegranate arils contain rich and nutritious proteins, antioxidants and probiotics, while refined arils become more vulnerable to oxidation and nutrient depletion. Pomegranate processing involves washing with sanitising agents to boost the elimination of microbial loads, pH changes, explicit antioxidants increases, and modified atmospheric packaging and temperature control (Bisht, 2020). The best results were obtained by washing the chlorinated arils with the "Mollar de Eche" cultivar with a mixture of ascorbium and citric acids, which were then stored in a polypropylene film at 1 C, which allows for the creation of a modified atmosphere suitable for the preservation of arils under very clean conditions. Sepulveda, et al) (such as honey and UV-C radiation).

Because many customers use printing and manufacturing materials in packaging, the option of packaging materials is made. Think of it as the consistency of the meal, its color and its nutritious composition. All this is accomplished by the influence of oxygen and light. Paperboard boxes and glass containers are used to pack recycled but low-density polyethylene items (LDPE). The disruption to the quality of the oxygen and light in the preparation was negative. The material used in the packaging of the package is also important for the creation of the capsule. PerezVicente et al. measured the impact of the pomegranate juice content on its color and bioactive components at 24'c ± 1' yr with a storage capacity of 50 percent /hectare, i.e. a 100 percent oxygen/oxygen atmosphere for 5 years. They argued that there was a change between the wrapping content and the organic quality of the juice.

At the moment, as long as there is no improvement in the purity of the food, the oxygen content is not affected. And pack the contents in a dry jar (which is the most damaging). Since red coloring in pomegranate juice can cause visual side effects, it is important to ensure that the light is at lower wavelengths and at higher color temperatures. Fruit juice bottles are found to contain more high-density polyethylene (HDPE) or polyvinyl chloride than other plastics used to retain organic compounds, in particular anthocyanins, vitamin C and other organoleptic substances. Light changes the viscosity of pomegranate juice resulting in a mechanism known as "aggregation" that is also thought to occur in blood plasma microcrystals (Behera et al., 2016). The findings of the study found that the content of anthocyanins in pomegranate in amber glass jars was slightly greater than in those contained in translucent bottles.

The more variables present, the quicker the color of the pomegranate juice can be changed. Due to various conditions, such as temperature, oxygen, light, pH, and enzyme activity, granate juice produces a multitude of pigments. In principle, the propagation of bacteria is the largest of the viruses. Numerous experiments have shown that pomegranate juice can be stored in a cool box and a cold stock (about 5 ± 12 °C) at room temperature for several days. There's a downside, though. Higher anthocyanin retention and lower pomegranate juice enzyme production have been reported in the cold storage of pomegranate juice. Report Byre, H., et al. "The development of organoleptic score for evaluation of pomegranate juice crushed and packaged by cold storage" Food Research Journal, Volume 41, Issue 6, pages 943-952, 2013, stated that pomegranate juice (level of organoleptic scores) which has been pressurized manually or mechanically under sufficient cold storage and then packed in glass bottles for up to 3 months can be processed satisfactorily under conditions of high organoleptic quality(Boyd et al., 2007). These forms of goods use a minimum of TSS, acidity, pH, maximum starch, carbohydrates and tannins when pasteurized at 70°C and 500 ppm sodium benzoate and are processed at room temperature for up to 60 days. Researchers have used these low temperatures to find out more about how they work. When they had an area to grow the bacteria, they gradually changed the climate to one that was straightforward and that helped the bacteria. Dyeing was faster at higher temperatures, especially at 35°C. (Environment). Looking at the findings over three months, Shelar and Sandhan found that the overall sugar levels had risen at higher temperatures of 25 and 30 degrees Celsius, then cooled off, and that the sugar levels had decreased when the temperature was held down to 15 degrees Celsius.

4.3.2 Passion Fruit

The tropics in the United States are endemically favoring the growth of these fruits. Passion fruits have a distinctive taste and scent and have thousands of uses for fresh goods as well. Some of the vitamins present in passion fruit are low in content and suitable for low-quality vitamins such as sound provitamin A, ascorbic acid, riboflavin and niacin. As sliced in two, the fruit pulp is discarded and then added to fruit salads, ice-cream or fruit juice (real fruit drinks). Many people think passion fruit to be a perfect culinary treat. Some have turned them into drinks and desserts that have some flavor, but whether they're going to taste is up to the individual's taste buds, and the name would depend on the owner (Chakraborty et al., 2011).

Figure 4.3: Passion fruit is known to produce passion juice.

Source: *http://www.freshexotic.co/products/fruits/*

Processing

Passion fruits are one of the most quickly fried fruits (including some cherries, not too hard to chew) and eat extremely easily as a juicy pulp instead of a full-grain pulp. There are many yellow-and purple-colored cultivars of Passiflora, but there are also a lot of people who buy these colors in order, for the most part, to ensure a more medicinal variety (Nagy, et al., 1990). Although the additives in wine are usually diluted, the acidity in red wine is between 3.0 and 5.0. White wines are most commonly filtered. Simultaneous commercialization of fruit production in Hawaii and Australia has resulted in a major separation of gelatin pulp from seed and rinder. In Hawaii, the spinning blades of the centrifugal extractor twisted through tiny holes in the moving centrifuge, where the pulp and seeds of the mechanically sewn fruit had passed through. By the Australian method, the fruit is scattered among the spinning disks, dropping the rind unattached along with the heart. Roller compression systems have been developed only lately, only in the last 5 years (Challies & Murray, 2011). The juice of the fruit is separated from the seed or pulp of the passion fruit by a thickness of 0.8-0.3 mm.

Few other devices have also been found with the development of some new devices for the removal and filling of the inner pulp half of the fruit. The single pressure of the yellow passion fruit juice at 17+oBrix at 2,8+pp is too high to be used immediately. The juice is better processed and the method of extraction is by freezing it, with 1:3 or 4 being reconstituted with drinking water, with a ratio of 55 to 60 sugar parts per 100 juice parts. Purple cultivars or new varieties with low acidity and low sugar are needed for the supply of nectar. Pureeing, an amazingly quick process produces a material that has a smooth but yet thick consistency that does not require freezing. There are a few ways in which the fruit juice can be pasteurized so that the food can be taken at a certain temperature and stored there before the paste has been produced. In addition, a variety of fruit juices are often heated with such enzymes, resulting in a pasteurized substance that is specific mainly because of the significant amount of starch within it. Follow the material that has created a less fragile taste by working the commodity in the most delicate and feasible concentration, which eliminates all products able to attempt aseptic production.

Passion fruit is a vital ingredient in many mixed tropical fruit drinks thanks to its sunshine and is well-known for its unheated (fresh or frozen) flavor. They have a decent efficiency without heating. Due to passion fruit, low or short-lived evaporators-centrifugal or drop-film evaporators will have enough passion fruit concentrate greater than 60 Brix. The flavor and the spice will certainly be back after the water has been purged (but do not match). This concentrate is very stable, can be stored at high temperatures of about 20°C and is used in tropical drinks such as juice mixtures, distilled and frozen in 200L barrels. There is no big pectin-producing industry, but the seed content is very high and the seed itself is a source of protein and oil. There are a lot of things that need to be done in the light of the recent research and creation of artificial honey (Crush & Frayne, 2011). Studies and journal papers have suggested that agriculture has not so far suffered from economies of scale and mixed crops. In addition, several other distinct generations of Passiflora are worth growing.

A foreign taste can be very surprising, for example, in the case of this exotic fruit. Passion fruit juice is used as the strongest ingredient in a wide variety of blended cocktails and exotic beverages in very small amounts. To achieve the best possible results, Bertuzzi has also constructed dedicated juice extraction machines and extremely comprehensive juice handling techniques. The juice extraction machines do not cause any shortage of fresh products. Fresh fruits of love are also shipped to our hospital and packed into our stainless steel all-oxidizer. We suggest that the implementation of a bristle machine be used for the complete cleaning of fruit. Our dedicated pulping systems range from 800 to 5000 kg/hr (8 to 5 tons per hour) from Passy Pulp and Super Passy Pulp. By blowing the fruit, they wring the taste/juice out of the pulp. The different shapes of the flower tubes are suitable for narrow-purple flowers as well as for all types of yellow tube. To exclude plant material from the seeds, we sell our "Super Creamer" and "Jobo Creamer" extracts which are not as refined as this pure formula. A centrifuge can be used to reduce the volume of pulp and also to produce a gelatinous material. Flesh term is used to increase the concentration of low viscosity juices by preserving the organoleptic properties. A device capable of absorbing the scent can be installed. The sterilizer/pasteurizer can sterilize and pasteurize diluted or fresh passion fruit juice, depending on the precise application order. If the medication is to be filled in aseptic bags in batteries or boxes with the aid of Bertuzzi Aseptic Fillers, the bags and boxes will be sterilized at high temperatures. Both Bertuzzi lines are capable of compliance with international food safety standards and can be entirely cleaned using automatic cleaning and sorting or centralized, fully automated CIP systems (Chazovachii, 2012). The specific purpose of CipMatic Bertuzzi is to guarantee the optimal and effective CIP (conversion of lumped extraction) of fruit processing plants.

4.3.3 Figs

Figure 4.4: Figs are known to be consumed when dry.

Source: *https://www.smallfootprintfamily.com/brazilian-fig-torte-recipe*

Ficus carica is an Asian flowering plant species known as the common fig in the mulberry family. The fig, the root of the plant, is also called and is also an essential crop in the different countries where trade is cultivated. It has been grown and cultivated since ancient times and is now widely cultivated around the world as an ornamental plant, native to the Mediterranean and Western Asia. The plant has been naturalized in scattered areas in Asia and North America. *Ficus carica* is a gyno-decided tree or large bush that grows up to 7-10 meters with a smooth white bark. The scent leaves are 12 to 25 cm long (4.7 to 9.8 in.) and 10 to 18 cm thick (3.9 to 7.1 in) (three or five lobes). The fig fruit grows like a hollow fleshy organ called syconium, bordered internally by various unisexual flowers. The tiny flowers are blooming in this cup-like setting. Syconium is a type of reproductive fruit, botanical and in fructose, commonly known as a seed. Its inner surface is lined with small fig flowers and therefore small (true) single seed fruits (Chakraborty et al., 2011). A tiny opening or ostiole, visible in the center of the fruit, after which each fertilized ovule, one per flower in its ovary, grows into a seed, is a narrow passage that enables the adapted fig wasp (*Blastophagous psene*) to pollinate the flowers. These "semences" (single-seeded fruits) are mature in every fig. The food-ripe, mature, druple syconium stem becomes a fleshy, artificial fruit with several single seed fruits. The whole fig fruit is 3-5 cm long and sometimes turns brown or purple with a green base. *Ficus carica* laticifer cells are milk sap. The sap of the green regions is annoying to human skin.

Processing

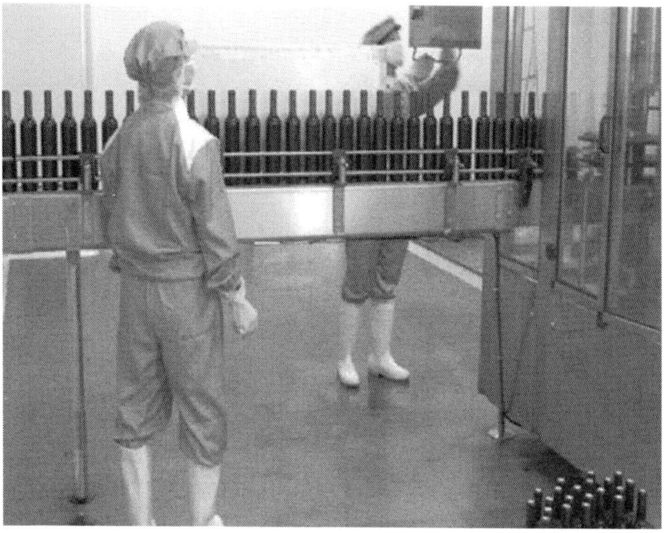

Figure 4.5: China is known to commercially produce fig juice.

Source: *https://genyondmachine.en.made-in-china.com/product/ZyQxvYrVh-BhJ/China-Figs-Fresh-Juice-Full-Automatic-Production-Line.html*

There is a method of drying where, by increasing humidity, figs are flattened, their pear-like shape disappears and spherical forms are formed. Natural figs can be distinguished on a consistent basis from where they were harvested: some are sun-dried or pressed, some are threaded on ropes or in chains. According to some, large batches of dried figs are sprayed with a mixture of water and salt, which then extracts the drying process and some of the vinegar (Dagar, 2014). When (literally, "when") figs are properly cooked, they show off their exquisite aesthetics and are very poetic. Since the figs have been cured, figs are the most commercially valuable. Patent applications for fig processing techniques have been made. The discovery involves the refining of fig juice and is part of the method of making improved fruit juices. A revolution has been made by: going by fruit collection, fruit purification, fruit carving, fruit heating and fruit destroying machines that can trap fruit and processed fruit juice, and then through the grinding process of juice, and juice with additional taste and flavor, and the machine with various kinds of additives and vitamins and the seedling machine that can be changed from the old one.

The method in which the juice is calibrated to the pH level of 3.5 and held at that pH level for 30 seconds, then to the pH level of 3.6 for two hours, and then to the pH level of the juice at the end of the two hours, is changed to 0.06 percent. It is a special-purpose clarification at night for 45-60 minutes. The innovation has easy use and fast setting and a lower price, and most fruit juices are like this. There are different methods used when using technologies in the processing of figs. In this pre-harvesting stage of *Ficus carica* , the juice pre-harvesting is characterized by the handling of fruit plucking, fruit harvesting, washing juice, antioxidants, measuring, heating, splitting and allotting; concrete fruit harvesting: fig fruit ripening requires adequate light and air. Heat the fruit with a strong enzyme (Dadabhau et al., 2013). Squeeze the fruit juice easily to 95 degrees Celsius and take thirty seconds. Note that the pH is going to reach to around 3.6. Add 0.06 percent of the volume of pectase. Trigger the reaction to commence at 50 degrees Celsius for two hours.

For thirty minutes, install a special fine agent. Using a clarification process such as CTA, after centrifuging the squeezed out juice and moving it through a certain clarification system, the juice extracted and the water are separated and then the solid-liquid in which the juice is removed. Then the juice is purified using filters. Pure or neutral fruit juice is produced after filtration. Measurement of 595 nanometers below wavelength Allocation: water purification and 70 percent allocation of white granulated sugar; formula: 30 percent clear fig extract, 6 percent white grain sugar, 3.0 percent maltose, 3.0 percent sodium cell, 0.03 percent acid, 0.35 grams/100-milliliter acid, 0.05 grams potassium sorbate, 0.0 percent ascorbic acid. The acid described in the paper would ideally be citric acid: fumaric acid: lactic acid: 4:1:0.01:0.5:0.3.

4.3.4 Papaya

The distinctive feature of the papaya fruit is that it is typically three to more than one hundred grams in varying lengths and that it can appear pale yellow to dark red-orange in color. The high pH of the flesh, which is most commonly in the range of 5.5 to almost 6 when eaten at maturity. The flesh is also a very good source of carotenoids. An example of its carotenoid nature is its reddish hue. The skin, flesh and meat of the papaya must be treated with unparalleled care in order to prevent the flavor of pawpaw healing. In our current processing method, Chan, Nagy et al. have been well established.

A surface steam preparation is a prepared material, the wrapping of which performs cutting and mechanical peeling in combination with a modified centrifugal peel separator, allowing the coagulation of latex and the hardening of the root flesh. In the finishing process, the skin is quickly discarded and chemically cleaned clean. It is important when stressed, that one absolutely eradicates the skin and the seed from the body (Davis et al., 2014). Papayas are a bunch of fruits; however, because foods have such high pHs, they would be pasteurized with citric acid (or combined with acidic juices) before being canned. The flesh is thermally modified by preventing the enzyme from conducting its operation. Because of its various applications, papaya root (or papain) is used for cooking. Papain is found in natural good strawberries and industrially in latex. It is not present in the ripe fruits, however. Use only high-quality sausages and fruit for the sake of high-quality meat or juice.

Another enzyme known as pectinesterase, which demethylates pectin, weakens the surface of pectin, making a less viscous gel, similar to gelatin. Avoiding freezing thus requires fast inactivation of the hot, sweetening or acidification to less than 3.6 pH. Acidification and aseptic processing are important methods for managing the consistency of bulk, the color of papaya, the soft taste of papaya, the high quality of carotenoids that make papaya popular for tropical drinks. To dilute the wine further, a pectic enzyme must be used to have an effect of about 3 times the concentration of up to 3% by volume. Finally, the Carica pentagon (Tepezcuintle) from the North of South America is worth recalling. The fruit looks similar to the orange blossom, but the fruit is more acidic, tastes much better, and is even more nutritious. The juice is highly popular in the region, but it is not well known abroad. In this study, researchers have been able to prove that species in a certain plant genus may have different tastes.

Processing

The period of maturation of the papaya fruit, during which the fruit undergoes limited processing, has been tested and the effect of the processing is determined on the physiology of the fruit. Cutting and polluting agents result in a spike in ethylene and drying. .The formation and softening of ethylene in the skin was also observed. The pulp where a piece of fruit is cut and sewn will experience greater growth and breathing compared to a piece of fruit that is more likely to be produced. In the early stages of ethylene fermentation and respiration, the fruit had more yellow tones and, once it was diced, seemed to harvest more ethylene, which proved to be better than most ripening methods. All is perfect, for the most part. When the berries were cut in half, they were removed. In one day, the production of ethylene glycol and breathing declined significantly at 4°C. It appeared as if the fruit was more than 50% poisonous, but the skin of the fruit was yellow and impossible to tear, and the inner fruit flesh was almost firm (Davis et al., 2014). The fruit is ideal for comparatively reduced processing in conjunction with lower storage temperatures (approx. 4 °C) to inhibit the development of yeast.

Cleaning of fruits. Fruits and vegetables also have microflora, much of which is also attributed to foodborne disease outbreaks. In consideration of the fact that fruit and vegetables are primarily eaten uncooked or uncooked (such as ready-to-eat sausages), microbiological protection is a vital factor in reducing the risk to consumers. Several clusters of fresh fruit and vegetable items have recently been attributed to poor health conditions. Appropriately disinfection is currently considered to be one of the most critical in allowing new fruits and vegetables to survive as they are so delicate. This process of sterilization and chemical modification impacts the quality, protection and shelf life of the end product. Even sanitization, which is the destruction of infected microorganisms after cleaning. The use of mechanical washing involves the use of chemicals to dry the cloth and then more chemicals to clean the tissue. The feasibility of preserving fruit freshness effectively by the use of sanitizing additives is encouraging, as is the case with the maintenance of freshwater, enriched by microorganisms. Sanitizers that sterilize bacteria can mitigate infection.

Sort and classification: Fruit and vegetables are sorted directly after they have been washed to produce a uniformly sized package by scale, shape, weight or color. The scale of the items will be used to determine the amount of drying and heating necessary and how much time will be needed (Davis et al., 2014). Classification play an important part in assessing the accuracy of a product. In conventional packaging inventory and labeling schemes, workers manually sort and label products that are not yet ready for distribution. This manual scale is labor-intensive, time-consuming and has no visible visual observation marks. The judgment of that decision suffers from ambiguity and oddity. New systems need to be improved in such a way that sorting and measuring are more efficiently minimized, precise and reliable.

Every kind of fruit must be peeled apart from being eaten for further processing. This can be achieved very quickly with a sharp, stainless steel blade. The knife must be made of stainless steel so that the cut it provides does not cause damage to the tissues of the plant. As part of food production, the peeling of agricultural crops is an important step in which the bulk of agricultural crops must be peeled. Peeling our items brings out the delicious aspect of the fruit (peel, nuts, and stalk). However, the susceptibility to spoilage increases due to the speed of the digestion process due to the time at which the microorganisms enter the tissues and the proximity of the tissues to the microorganisms. The boss of the mine. Cutting is needed to complete the heating, drying and packaging process for standard parts. Different fruit is classified into three different types, strips, loops or shreds. To deter microorganisms from eating the food we eat, we must ensure that our utensils are sharp and sterile. To minimize the volume of ground-up beef, non-thermal processing involves a significant amount of energy. This approach sets the tone for a profitable undertaking (Das & Prakash, 2009). Although the extracts are made by cutting uncut fruit pieces and treating them with crème anglaise. However, damaged tissue is typically subjected to alcohol-based juice production, which promotes microbial growth. By making incisions in fruit tissues, the fruit will be able to retain more calories, more oxygen and longer shelf life. The cutting method results in the aggregation of fluids on the cut surface, which increases microbial loads and enzyme production.

Blanching is a cooking process in which the food content is subjected to fast boiling (10-15 seconds) to inactivate the enzyme and (1 minute) to tannin. The color changes; the texture lacks its original properties; the taste and nutrient quality of the texture reduces and thus the atmosphere is released. The biggest advantage of whitening before canning is that it helps to decrease gas pressure during filling and decreases the amount of gas entering the can. Because all enzyme activity is already inactivated at the end of the process, all remaining continuous reactions are not needed. When vegetables are aged to lose their color, they are easily cooled by being sprayed with cold water or moved to a cold water flow as they pass to the next step of the product processing. There are refrigerators with a hot air evaporator fan, which cools the air and the evaporator, and also holds the water inside the refrigerator. There is a way where water is bleached to make it more alkaline. It needs hot water to be used. However, low-temperature blanking (LTLT) and low-temperature fringing (LTF) can delay or impact crop growth and nutrient uptake.

As well as high-temperature bleaching and shorter length versions for LTLT and HTST. Our water treatments also have the advantage of being in a more uniform shape, which results in lower temperature treatments. Steam bleaching involves the circulation of fluid into a pressurized, fast-moving, whirlpool-like chamber that has a dramatic effect on algal biomass. Before entering the tub, the temperature is usually pre-determined and the steam flow rate is monitored in the headroom. As an optional technique, adding a certain portion of the liquid with the remaining part can shorten the heating time and make the food softer. As a result of advances in microwave washing, low water usage and improved articulation of nanomaterials have made it possible. To selectively remove gas, hydrogen chloride from lignosulfonate used in the manufacture of paper sheets, heat gas bleaching was tested to increase the water content and effectively avoid dehydration (Dixon et al., 2001). As this approach eliminates the after-effects of "actinic" whitening additives, the drawback of bleaching is that the use of the same ingredients found with the traditional alternative reduces the use of the products. This method of treatment is still very advanced and needs more study. The papaya undergoes the entire process and it is canned afterward as a method of storage.

4.3.5 Custard Apple

The yellowish-green fruit of the Annona plant family, is the Custard apple. There has not yet been a well-developed post-harvest process for these fruits, with some handling problems. Rapid fruit softening following harvesting is a major ongoing issue, especially in transport and marketing. The custard apple fruit can be improved for up to 12 days when the atmosphere is changed to 10 °C while the pulp can be processed for six months with potassium metabisulfite. Nutrient potential can be improved and new flavors/samples can be given across a range of value-added ingredients, including ready-to-serve beverages, fermented beers, ice cream, squash and toffee. The purpose of this study is therefore to clarify some of the important findings on storage, in particular after harvesting, the addition of processing value and the storage of this fruit. It is typically favored and added during the later stages of preparation, and appears to crack slightly when in contact with the heat. The custard apples can be frozen and stored for another day for some time after they have been picked. It can be processed effectively at cold temperatures for up to 9 weeks with a RH of less than 85 percent at 7-10 °C. Colder temperatures in storage is what triggers the cooling. Due to its composition, the pulp can be frozen for use in the ice cream industry (Fakir et al., 2018). A circular hair comb was developed at CRIDA as a simple technique for extracting the soft apple pulp from the scooped fruit of a stainless steel sieve. This pulped refill will be sold by the ice-cream company. It is necessary to restrict the processing of the pulp to produce soft fruits.

Chapter

5

MEDICINAL AND AROMATIC PLANTS

CONTENTS

5.1 INTRODUCTION

Medicinal plants and aromatic plants were discovered during prehistoric times. Most aromatic plants have medicinal properties. Such plants include garlic, ginger, onions, and chamomile. They were important and are still important because they produced chemical substances that are responsible for curing illnesses as well as fighting pests. Over the years, plants with a lot of useful chemical compounds have been studied in detail. It has been established that only specific parts of the plants can be used in the healing process (Galán, 2008). A lot of scientists are working towards ensuring that the medicinal herbs are safe to use and that they do not contain a lot of toxins.

The Sumerians were the first people who studied and used these medicinal herbs, opium was among the many medicinal plants that the Sumerians used. Dioscorides was a doctor from Greek and he worked for the Roman army, he is well known because he identified over 1000 recipes for making medicinal herbs while using over 600 plants that had medicinal value. A lot of medicinal plants have since been studied and a lot of medicines have been introduced to people and they include, aspirin which acts as a pain reliever, quinine that is used to treat malaria and opium.

These types of medicines are used in areas that are not highly industrialized, the reason being these medicines are not expensive and they can be accessed easily than Western Medicine. Statistically, in the year 2012, the medicinal plants were valued to be worth 2.2 Billion dollars. The World Health Organization has been set up to come up with safety measures that safeguard the people who use these medicinal plants. It also encourages the use of medicinal herbs. There are problems that are faced by these medicinal plants that are the adverse climatic conditions, the destruction of their natural habitat and also the risk of being over-harvested in order for them to be able to meet the demand of the people.

5.2 ALOE VERA

Aloe barbadensis is the scientific name for Aloe Vera. It is well known for its therapeutic characteristics as well as the fact that it has other benefits when it comes to the health of a person. It grows in dry areas as well as tropical climates, especially in countries like Africa, Asia, Europe and also some parts of the United States of America. A long time ago the health benefits of this plant were discovered. It became popular among the people and everyone used to treat various diseases that ranged from skin issues to problems when it came to digesting food. These discoveries were made and have been used for over 6000 years. This plant was considered to be immoral and it was given to the pharaohs in Egypt as gifts in funerals (Gera et al., 2013). Over the past years, Aloe Vera has been used by people to treat wounds as well as preventing hair loss and treating issues that are related to digestion.

Figure 5.1. Aloe Vera plant and the health benefits.

Source: *https://www.healthline.com/health/7-amazing-uses-aloe-vera*

The following are the therapeutic characteristics of Aloe Vera:

5.2.1 Wound Healing

When it comes to the process of healing a wound, it occurs in three stages. The first stage involves the infiltration of leukocytes, inflammation and hyperemia. The second stage is where the tissues that are dead are removed and the last stage is where the removed tissue starts to regenerate and form new tissues.

Recently, studies have shown that Aloe Vera is used to treat people who have first to second-degree burns. The property of wound healing has been brought about by a certain compound that is found in Aloe Vera and it is called Mannose-6-phosphate. This plant is also very important when it comes to decreasing the levels of glucose in the body and as a result, the healing process of the wound is sped up, especially in people who have diabetes (Gururaj et al., 2017).

It helps in the healing process of wounds because it helps to reduce inflammation of the wound which helps in the healing process, this whole process helps to form a scar. Certain compounds help in the healing process that are found in Aloe Vera and they include, non-flavonoid polyphenols compounds phytosterols and indoles which are also beneficial when it comes to protecting the wounds against infections.

5.2.2 Healing Canker Sores

A lot of people at different times of their lives usually get ulcers in their mouth as well as canker sores. Some of these sores occur in the mouth and usually take about less than a week for them to clear up. Often these sores are usually very painful and can cause a lot of discomfort. Aloe Vera helps to heal these sores as well as clear them up. Studies have shown that aloe vera helps to reduce and heal the sores. One is required to apply aloe vera to the affected area. Aloe Vera also acts as a pain reliever when it comes to the mouth ulcer as well as the canker sores.

5.2.3 Treating Constipation.

Studies have shown that aloe Vera juice helps to cure irritable bowel syndrome, which is abbreviated as IBS. Some of the symptoms that a person may experience when they have irritable bowel syndrome are constipation and diarrhea. The symptoms can be cured by taking aloe vera juice. This juice is mainly important for helping to cure and treat the problem of constipation (Haque et al., 2018)

The juice helps to reduce the inflammation of the intestine that is responsible for causing this syndrome. Aloe Vera as we have seen contains anti-inflammatory characteristics. People who have this disease usually treat the symptoms that are accompanied by this disease by consuming a diet that does not contain a lot of carbohydrates and instead they take food that is rich in fiber that helps in the digestion process.

The compounds that are present in aloe vera are able to treat the problem of constipation. As soon as one consumes aloe vera juice, after 6 hours the problem of constipation is completely cured.

5.2.4 Anti-Tumor Activity

There are certain compounds found in aloe vera known as glycoproteins and studies have shown that these compounds can be able to cure tumors as well as ulcers and also they can increase the proliferation of the dermal cells. A lot of research has been done for them to back up this theory.

These proteins help to treat tumors that are often cancerous. Other compounds play a very important role in ensuring that stomach ulcers and other types of ulcers are treated and cured.

Aloe Vera also helps to prevent the tumors from growing and also developing. It is incorporated into the actual treatment and studies have shown that it plays a major role in actually helping to cure these two problems.

5.2.5 Healing of Fungal Infections

Aloe Vera helps to treat and cure some fungal infections. They contain compounds that are important in the healing process of fungal infections. Some of these fungal infections include: dandruff, athlete's foot and other fungal infections that affect the skin like ringworms.

The Aloe Vera is applied to the affected area as a medicine and the healing process begins. Some medicines have been manufactured with aloe Vera as an ingredient. It also helps to clear acne from the skin and it is highly recommended by a lot of physicians (Hoang, 2008). It also contains compounds that act anti-aging properties. It leaves the skin feeling supple because of its hydrating property and completely cures and treats acne.

5.3 CHAMOMILE

When the chamomile leaves are crushed, they produce certain compounds known as terpenoids and flavonoids which have medicinal property. This plant is used to cure different diseases in human beings such as ulcers, insomnia, rheumatic pain, menstrual disorders such as menstrual cramps and hemorrhoids. The oils that are obtained from these plants are fused to make cosmetic products as well as they are used in aromatherapy. There are many ways that this plant can be consumed and one of the most common ways is through taking it as tea. There are a lot of studies that have shown that this plant has healing properties. It is also highly recommended by a lot of physicians. It can also be dried and the powder can be used to heal various diseases. Traditionally this herbal medicine was used to target diseases as well as heal wounds and also it was used as a painkiller. It is also used to treat other diseases that are related to the stomach which include diarrhea, being nauseated, vomiting as well as indigestion.

Figure 5.2. Chamomile plant. The history as well as the health benefits.

Source: *https://en.m.wikipedia.org/wiki/Chamomile*

The following are the therapeutic characteristics of this medicinal plant:

5.3.1 Treatment of Common Cold

A lot of people are usually prone to the common cold, which is usually caused by a virus. It is usually not a serious disease but it only affects the upper parts of the respiratory tract. This disease is usually not a death sentence but it can develop to become pneumonia which can be life-threatening if left untreated. Chamomile is very helpful when it comes to curing a common cold as well as curing pneumonia. For it to work one needs to boil the chamomile plant and inhale the steam that is produced because this thing is what usually cures these respiratory diseases (Hossain et al., 2017). So far this plant has proven to be very efficient when it comes to the curing of the common cold and other respiratory diseases.

5.3.2 Treating Eczema

Some creams are usually made out of chamomile that help to treat and cure the problem of eczema, specifically atopic eczema. According to the research made, the effectiveness of this herbal medicine in the treatment of eczema is approximately 60%. Some creams only contain the Roman chamomile that is very efficient when it comes to helping a person is his or her discomfort that is usually accompanied when one has eczema, the screen is usually applied directly to the affected areas. This type of chamomile is usually very rich in other compounds that are useful when it comes to the treatment of allergies. This plant is used to treat eczema and also to manage this problem, it has proven to be of value when it comes to other skin conditions. More research is yet to be done to determine how effective this plant is when it comes to the treatment of eczema, so far people have not recorded any adverse reaction to this herbal medicine.

5.3.3 Treatment of Hemorrhoids

A lot of research has been made about the effectiveness of chamomile in the treatment and the management of hemorrhoids. It has been proven that some of the compounds that are present in chamomile are responsible for treating and managing as well as preventing the risk recurrence of hemorrhoids. the specific compounds that are present in chamomile that is used to treat hemorrhoids is known as tincture of chamomile, this compound is effective in reducing the inflammation that is caused by hemorrhoids and as a result, the hemorrhoids are completely eliminated from one's body. It is a process but chamomile has proven to be very effective when it comes to the whole process of dealing with hemorrhoids (Hilson, 2016). One needs to incorporate this medicinal herb in their bathing water and use this water to wash the affected area.

5.3.4 Treatment of Sleeping Disorders

When it comes to the traditional settings chamomile was used to induce sleep as well as cause a calming effect on people. Chamomile was taken as well as it was infused in oil that was used during aromatherapy. It was very effective and it is still very effective in treating insomnia among people. It is used to tranquilize people as well as make them sleep better and faster for those people who find it hard to sleep at night. The compound that is responsible for the tranquilizing effect is known as flavonoid which binds to some of the receptors that are found in the brain, these receptors are known as benzodiazepine. According to research, it has clearly shown that people who take this chamomile tea usually sleep immediately for a period of 90 minutes (Islam et al., 2019). The essential oil that is used in aromatherapy is responsible for reducing stress when it is inhaled. It is therefore very evident that chamomile is used to cure sleeping disorders as well as reduce stress because of its calming effect.

5.3.5 Wound Healing

A trial was conducted where 14 patients were made to undergo the process of getting a tattoo and afterward chamomile was used on the tattoos. It was very efficient, especially during the tattoo as well as speeding up the healing process of a tattoo, it was therefore agreed that chamomile was very effective in the wound healing process Those people that used chamomile after getting their tattoos, the rate of recovery was higher than the rest of the people. Chamomile is actually very efficient in the process of wound healing as compared to corticosteroids. It was seen that the ones that were associated with the tattoo decreased fast when chamomile was used. A lot of studies and research are still being undertaken to maximize the efficiency of chamomile in the healing process of wounds.

5.3.6 Curing Vaginitis in Women

This problem is usually very common in women and it is usually an infection that is accompanied by itchiness discharge from the vagina that is not normal as well as pain while urinating. This disease does not affect people of a certain age instead it is a problem that all women may experience at some certain point in their lives. As much as this disease affects or women it is most common to women who are nearing the menopause stage as well as those who are already experiencing menopause. It is mainly caused by a reduction in the level of estrogen in women. Douching using chamomile might be very effective when it comes to the treatment and healing of vaginal inflammation (Jamnadass et al., 2011). This might still be accompanied by some undesirable side effects. Research is still ongoing when it comes to the function of chamomile in the treatment of vaginal inflammation.

5.3.7 Treatment of Sore Throat

Chamomile acts as a lubricant when it comes to the throat; this helped to prevent and cure sore throat as well as hoarseness. This whole research involved giving a group of people purple chamomile extract, these people had undergone a series of different operations. It was given before the operation to lubricate the trachea before integration and the other group was not given any form of lubrication before intubation. These groups of people were then given or rather they were put under general anesthesia. Those who received chamomile did not experience any sore throat after the surgery but those who did not receive the chamomile puffs extract reported cases of sore throat. It was very evident that chamomile played a very important role when it came to the lubrication for the throat as well as ensuring that it was conclusive to say that chamomile can in fact help to cure and treat sore throat.

5.3.8 Treatment of Anxiety and Seizures

This medicinal herb has been researched and it has been concluded that it is used to treat generalized anxiety disorder which is abbreviated as GAD. It contains different compounds that have been related or rather have shown the capability of reducing anxiety as well as treating seizures. The components include picrotoxin that is mainly used by seizure patients and apigenin, which has proven to reduce convulsions in seizure patients as well as reduce anxiety (Kotwal et al., 2011). Continuous usage of chamomile which is often incorporated in tea helps to treat and reduce the cases of anxiety as well as it is used in the treatment therapy of seizures. More research is still ongoing when it comes to the treatment of anxiety and seizures the use of chamomile.

5.4 GINGKO

This plant is highly known for the medicinal value of its leaves. This plant is very common to the Chinese people and it has a lot of health benefits. It is often called a living fossil because it started being cultivated a long time ago. In China, people usually use these plants' leaves and seeds as a form of medicine but modern research has revealed that even the leaves have a lot of medicinal value. This tree is usually very important in dealing with issues that affect the brain and also the circulation of blood in the body.

Figure 5.3. A representation of the ginkgo plant, the health benefits and therapeutic characteristics

Source: *https://www.healthline.com/health/ginkgo-biloba-benefits*

The following are the therapeutic characteristics of this plant:

5.4.1 Treating Depression

During the research done on animals, ginkgo when incorporated into the treatment therapy of a person who has depression then he or she might be cured of depression. It performs this function by ensuring that it treats all the symptoms that are related to depression and as a result, the person receives healing or rather he or she is cured.

This research involved the use of mice, two different groups of mice that will underwent a very stressful situation. One of the groups was given ginkgo and the other one was not, the results were, the group that was given ginkgo seemed to handle the pressure from the stressful situation very well and it did not affect them emotionally on the other hand the other group did not handle the pressure very well and it affected them emotionally (Koohafkan & Altierim, 2011).

Ginkgo is well known for its anti-inflammatory nature, this helps a person to handle a stressful situation but one that has a lot of pressure calmly because ginkgo helps the body to adapt when a person is under stress by reducing the level of hormones that cause stress.

It is conclusive to say that ginkgo can be able to treat as well as cure depression but a lot of research is still ongoing to prove this theory. It is recommended by doctors as an antidepressant to those people who are undergoing serious cases of depression.

5.4.2 Treatment of Headaches and Migraines

Traditionally, the Chinese people used this medicinal herb to treat headaches and migraines. It can treat the different types of headaches depending on what caused the headache as well as the migraine. Due to its nature of reducing inflammation as well as the fact that it is an antioxidant in nature, as we have seen ginkgo plays a very major role in reducing stress levels in a person. If the headache is caused by stress or rather a stressful situation, then ginkgo can be used to treat the headache or the migraine.

Ginkgo has another property which is, it is very useful when it comes to dilating the blood vessels that are constricted and as a result, if the headache is caused by a reduction in the amount of blood that is flowing to the that area as well as blood vessels that are constricted then ginkgo can treat this type of headache and migraines.

According to a lot of research by different scientists, there are very many more risks that are associated with the use of ginkgo in the treatment of headaches and migraines

5.4.3 Treatment of Sexual Dysfunction

This medicinal herb can be used to treat many issues that are related to the sexual dysfunction of both females and males. It can treat erectile dysfunction in men as well as treat low libido in both women and men. Ginkgo has a certain property that involves increase nitric oxide in blood levels, which in turn helps to circulate the blood when the blood vessels are dilated (Kalinganire et al., 2007).

There are some medicines made from ginkgo that are meant for curing depression in people and research has shown that they may be effective in treating sexual dysfunction, ginkgo has been more successful in treating sexual dysfunction than any other herbal medicine.

The use of ginkgo to increase libido in women as well as sexual satisfaction in women who are undergoing therapy about their sex lives is still ongoing.

The use of this medicine plant over a long time, has proven to be very effective in the treatment of sexual dysfunction in both women and men. Like many other medicinal herbs, this function still needs to undergo a series of tests before it is made a permanent solution in the treatment of sexual dysfunction.

5.4.4 Helps to treat Various Cases of Inflammation

The term inflammation means, the body's general reaction when a certain part of the body has been injured or when something foreign has invaded a particular part of the body. This inflammation could be a result of some diseases as well as injuries that the body is currently undergoing. If inflammation is left untreated it can cause irreversible effects to the tissues as well as the DNA of the body. Ginkgo has an anti-inflammatory property that helps in reducing the inflammation that has been caused by these diseases and injuries (Krishna et al., 2019). Some of the diseases that cause inflammation and where ginkgo has shown to treat this inflammation include: cases of cancer, diseases related to the heart, stroke, arthritis as well as irritable bowel disease.

According to the research that has already been made by different scientists all over the world, it is conclusive to say that ginkgo plays a very important role in the treatment of inflammation caused by certain diseases and injuries that when left untreated have proven to be life-threatening.

5.5 ECHINACEA

This medicinal herb is grown in Canada and Europe as well as the Western states of the United States of America. The parts of this plant that have medicinal value include the leaves, the flowers as well as the roots. It was commonly used in the past by the people of Indian tribes. As a result, many people adopted the use of these medicinal plants in the treatment and healing of different diseases. This herbal medicine was used by people of the United States and it was very popular among people because could cure different bacterial infections, it was used as an antibiotic by different people.

There are different forms in which these medicinal plants can be consumed and they include it can be in the form of tablets, the medicine can be taken in the form of juice as well as it can be incorporated in tea. This medicinal herb acts by activating body chemicals which are very essential in the treatment of cold and flu. According to the Research made it is also safe to assume that this medicinal herb contains some compounds that are very effective in treating yeast-related diseases as well as treating diseases that are a result of fungal pathogensis.

Figure 5.4. A representation of Echinacea plants and their therapeutic characteristics

Source: *https://www.medicalnewstoday.com/articles/252684*

The following are some of the therapeutic characteristics and the diseases that this medicinal plant is able to treat.

5.5.1 Treatment of Common Cold

When one takes this medicinal herb when they are experiencing symptoms of cold, then it is possible for them to be healed and it is all thanks to the compounds that are found in this medicinal herb. A lot of research has also shown that Echinacea helps to prevent the common cold as well as flu (Marín et al., 2008). This is because of its anti-inflammatory property that helps to fight off the viruses that result from the occurrence common cold.

5.5.2 Treatment of Eczema

Certain creams have been made out of an Echinacea medicinal plant. These creams are applied directly to the affected areas and it has been established that after 12 weeks of consistent use, these creams can treat the symptoms that are accompanied by eczema such as, the affected area being itchy, with a swelling, and red in appearance as well as the area being dry. Even though it takes a lot of time and patience for these creams to work, Echinacea has proven to be among the best treatments that can be used to treat problems of the skin like eczema. The compounds that are found in Echinacea are responsible for treating eczema. More research is still to be done on the complete effectiveness of this medicinal herb in the treatment of eczema and also the treatment of symptoms that are accompanied by eczema that proved to cause discomfort to the affected person.

5.5.3 Treatment of Mild forms of Gum Diseases such as Gingivitis

There are certain more trees that are found in different pharmaceuticals that contain Echinacea as well as other medicinal components that help in the treatment of gingivitis. For the treatment to be successful one needs to use the mouth rinse for 14 days and he or she should use the mouth rinse three times a day that is, in the morning, in the afternoon as well as in the evening (Pingali, 2007). The continuous use of this mouth rinse that contains the Echinacea medicinal compound has shown great promise in the treatment of gum diseases but this whole process needs to undergo more research to make it a permanent solution in the treatment of gingivitis, which is a gum disease. In the treatment of gum diseases like gingivitis, one can either choose to use the mouth rinse or use a specific patch that is meant for the mouth, specifically the gums.

5.5.4 Treatment of Swelling of the Tonsils

In this case, two medicinal plants are used together in the treatment of tonsils as well as other issues that are related to the throat. These two medicinal herbs include Echinacea as well as sage. For the treatment of tonsils as well as other throat-related issues, one needs to spray the mixture of this to medicinal herbs in their mouth for five days and after every 2 hours: 10 times a day. This helps to reduce the inflammation that is caused by the swelling of tonsillitis. It can also be consumed as drops that contain other components like, Esberitox, schaper and brummer which are supposed to be taken for a period of two weeks and three times a day that is in the morning, in the afternoon and in the evening.

These two methods have proved to be very effective in the treatment of tonsilitis as well as a sore throat. More research is being conducted on the effectiveness of these two herbal medicines in the reduction of inflammation that is caused by tonsils.

5.5.5 Treatment of the Inflammation of the Eyes

A lot of research has been made on the effectiveness of Echinacea in reducing the inflammation of uveitis. Doctors recommend that a person take 150 milligrams of medicine made out of Echinacea which has the following components, iridium, SOOFT Italia SpA. One is required to take this medicine two times a day for 4weeks accompanied by eye drops and steroids that treat eye inflammation (Sarker et al., 2017). Note that this medicine is not responsible for improving the vision of a person; it is only effective in reducing the inflammation of the eyes.

A lot of research is still being done in order to establish if Echinacea can be used to treat various diseases such as malaria, migraine headaches, syphilis, typhoid, urinary tract infections, and yeast infection as well as if it can be used in the treatment of attention deficit hyperactivity disorder that is abbreviated as ADHD.

5.6 PRIMROSE OIL

This medicinal herb is mainly known for treating conditions like eczema as well as pre-menstruation syndrome. A lot of research has been done with this medicinal herb and it has been found that it contains anti-inflammatory characteristics that help with conditions like diabetes, neuropathy as well as atopic dermatitis. It has also shown very promising functions such as treating breast pain. According to the research that has been made just recently it is clear that the evening primrose oil improves the quality of people who live with multiple sclerosis as well as other people who deal with issues related to hormones and insulin. More research is being done to find other benefits and other diseases that these medicinal plants can be used to cure (Srivastava & Sarkar, 2017).

Figure 5.5. The evening primrose oil, the dosage and the benefits.

Source: *https://oilswelove.com/6-best-evening-primrose-oil-supplements-soft-gel-dosage-and-benefits/*

The following are some of the diseases that it cures.

5.6.1 Treatment of Acne

It helps in the treatment of acne by reducing the inflammations on the skin as well as treating the cuts that are on the skin. It also keep the skin hydrated because it helps to retain the moisture of the skin. The evening primrose in the research done in 2014 helps in the treatment of cheilitis that is a condition whereby the lips of a person become inflamed and one experiences pain in the lips. It is caused by a medicine that is used to treat acne that is known as Accutane and it has a compound known as isotretinoin.

More research that was conducted showed and proved that the supplements of GLA help in the treatment of acne cuts and bruises that were inflammatory and non-inflammatory. This medicinal herb is taken thrice a day that is in the morning, in the afternoon as well as in the evening for 8 weeks and in six 450mg capsules.

5.6.2 Treatment of Premenstrual Syndrome

This medicinal herb is very efficient when it comes to the treatment of symptoms that are accompanied by premenstrual syndrome. These symptoms include: irritability, depression as well as bloating.

A lot of women before their monthly cycle experience a lot of discomforting symptoms which are caused by increased levels of prolactin in which case most women are usually very sensitive (Sahoo et al., 2019).

The evening primrose oil was very beneficial when it came to relieving and treating the symptoms that are accompanied by premenstrual syndrome. It is taken four times a day for 10 months and one should take approximately 6 to 12 capsules.

5.6.3 Treatment of Heart Diseases

A lot of people who suffer from heart diseases usually turn to medicinal herbs such as Evening primrose oil which contains compounds that play an important role in treating heart diseases. Statistically speaking, heart diseases are the cause for a lot of deaths in the United States of America whereby it is estimated that 600000 people succumb to these illnesses of the heart.

The research to determine whether evening primrose oil was efficient in treating heart diseases was done on rats in the year 2014. The evening primrose oil helps to reduce inflammation of various parts of the body like the heart. It is also important in reducing the cholesterol levels in the blood. The inflammation of the heart is what causes the diseases that are related to the heart.

This medicinal herb while treating heart diseases should strictly be taken under the supervision of a medical practitioner. One should take the medicines for four months in 30ml-10 doses of evening primrose oil. It should also be noted that if one is under other medications that treat heart diseases then he or she ought to be very careful while taking this medicinal herb (Sayer et al., 2012).

5.6.4 Treatment of Nerve Pain as well as Bone Pain

The condition that causes pain in the nerves is known as peripheral neuropathy and it is usually a result of diabetes and other health-related issues. There is a certain component present in evening primrose that is known as linoleic acid and it plays a major role especially when it comes to treating the symptoms that are caused by peripheral neuropathy. These symptoms include: sensitivity to high temperatures as well as low temperatures, loss of the sense of feeling in a particular area of the body where nerves are located, general body weakness as well as tingling sensations in the body that are a result of numbness.

For the treatment of peripheral neuropathy, one is required to take the capsules that contain evening primrose for approximately one year as well as ensuring that the capsules are not of more than 480mg.

The issue of bone pain is usually caused by a certain illness known as rheumatoid arthritis that is often caused by inflammation and is mostly considered to be a chronic disorder. According to the research that was conducted in the year 2011, the component that is known as GLA that is usually found in the evening primrose oil is the component that is responsible for treating arthritis that affects the bones.

For one to ensure that they are completely healed from the chronic pain and also to ensure that they have properly managed the pain then it is advisable, for one to take the capsules that contain the evening primrose oil for a period that does not exceed 12 months and they should not exceed a maximum dose of 6000mg.

5.7 GARLIC

Garlic is mostly referred to as onion and it is often confused with onions because they highly resemble each other. The scientific name for garlic is known as *Allium sativum.* This medicinal herbs been around for a long time and approximately for over 40 centuries thus making it the oldest medicinal plant to be grown (Shackleton et al., 2007). Although some believe it was discovered in Siberia, other people believed that the Asian people were among the first people to discover this plant meaning that this plant originated from the eastern parts of the world, especially the northern side. It was mostly used to treat issues related to the heart as well as issues that affected the blood therefore it was very popular when it came to the treatment of blood infections.

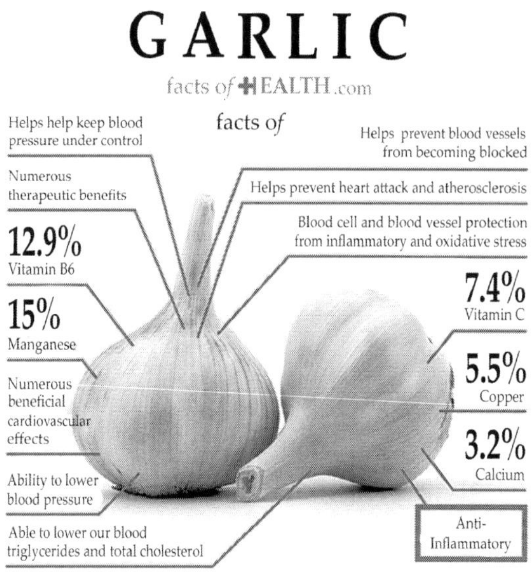

Figure 5.6. A representation of the garlic plant as well as the health benefits.

Source: *https://creeklife.com/blog/garlic-health-benefits/*

The following are some of the diseases that it is used to heal as well as treat.

5.7.1 Treatment of Cardio-Vascular Infections

Over the years, garlic has been growing in its popularity for its success in the treatments of cardiovascular diseases. Doctors and medical practitioners have always reported positive results from their researches conducted on this herb. Its use has grown over the ages and this has prompted more and more researches on this promising territory. Most of the heart-related diseases have all shown tremendous improvements to *diallyl trisulfide* which is an enzyme that is found in garlic. Most of the people, especially those with hypertension after being put on this revolutionary garlic supplement of 4 doses of 100mg showed very improved results after a very short period. Also, people who have heart attacks and have through cardiac arrests showed very positive results after being put on garlic treatments. They help in making the healing process after a heart surgery move very fast (Singh et al., 2014). The antioxidants in the garlic see to it that the blood pressure of the patient remains normal that in turn speeds up the healing process.

5.7.2 Treatment of Various Infections in the Body

Garlic has so many components and compounds which makes a perfect antibiotic that is recommended by a lot of doctors. Its ability to eliminate many infections that are caused by bacteria that may attack people especially those that are suffering from life-threatening and chronic diseases tend to be given medicines that have been made from garlic because it is well known for its ability to fight bacterial infections. Thanks to its natural nature too garlic are perfect for very many infections that tend to beat other inorganic antibiotics that are prescribed and therefore garlic has proven time and again to be a very worthy antibiotic.

5.7.3 Treatment of Various Parts of the Body that Improves the Physical Health of a Person

In life, very many people face various life-threatening and traumatic accidents. At times these people can survive barely with some very bad physical. According to the history of garlic and the Greek, garlic possessed some medicinal components that can improve a person's physical health by treating various issues that the person might have and might not be aware of. The Greeks used it as a means of boosting the physical abilities of their athletes. These same theories are applied today to help victims regain their physical abilities. These people are given some garlic-based medicines as well as concoctions which in turn boost their physical abilities and help them to keep up with their practices that help them eventually get on their feet and maybe eventually regain their strength back making them fully healed.

5.7.4 Treatment of Flu and Common Cold

Flu are some of the most random highly transmittable and most popular infections amongst people. They are simply common communicable diseases and therefore they must be treated efficiently with haste before they are capable of starting various pandemics that affect the respiratory tracts of people. Garlic is a major remedy for these sorts of diseases and it is also very important when it comes to relieving the symptoms that are accompanied by a cold. This eventually gets to the flu and helps to heal it completely.

5.7.5 Treatment of Fungal Infections

Garlic is a very strong herb with a very scent that made it even be avoided by those people who were very rich in the past. This is as a result of the antioxidants and the different enzymes in the herb. The herb is divided naturally into sections known as cloves. These cloves were the ones that are needed for the whole fungi healing process. These fungi include even the more popular athletes' foot. You need to crush the cloves or pulp them and then with coconut oil to create a fluid that is responsible for the total and complete healing of all the fungi infections. Also, it's important to know that applying garlic directly to the affected can be quite corrosive and therefore it may cause itching and also the growth of pimples on the skin that may cause different skin diseases. Garlic has so many benefits to the medical sector and some others that we've not really looked at in-depth like, garlic can prevent some mental issues like dyslexia and loss of memory, it also affects helping to heal a liver that was previously damaged by excessive consumption of alcohol (Tiwari, et al., 2018).

5.8 GINGER

The scientific name of ginger is known as *Zingiber officinale*, which is a medicinal plant of the family *Zingiberaceae* that was first grown by the local people of Asia which meant that the people had some religious attachments to them. Also being a native Asian herb, it was among the first plant that was taken out of the Asian social settings to the rest of the world. It was a very valuable commodity of trade to quite a lot of people who regarded this plant that has a stem that is not actually a stem but a collection of rolled leaves with some yellow flowers. This herb like many other also has some medicinal effects that has a direct effect on the health of the patient. With the compound found in ginger, ginger becomes a very important medicinal herb. Ginger can also be used as a spice because of the aroma and taste that it adds to food. It is used in many countries across the world.

Figure 5.7. A representation of the ginger plant and some of its benefits.

Source: *https://slowaging.org/health-benefits-ginger/*

The following are some therapeutic characteristics of ginger and the diseases they heal.

5.8.1 Treatment of diabetes

For the longest time ginger has been used to help lower the blood sugars in the body. Tests conducted on people that are type 2 diabetic concluded that diabetes can be reduced significantly by taking ginger doses as prescribed by the physicians. The reduction of blood sugars is of very valuable importance especially on people suffering from diabetes. They are in constant fear of developing heart diseases and even in extreme cases having cardiac arrests as a result of these sudden increases and decreases in blood sugar levels. Therefore, physicians end up giving a prescription of organic powder ginger to help keep the blood sugars in check (Hossain et al., 2017). The use of ginger ,therefore, helps reduce the fatalities of diabetes and also keeps the already infected from succumbing to the disease.

5.8.2 Healing of Joint Pains

Ginger boasts to be one the most resourceful organic ways of reducing inflammations. The consumption of raw, cooked or crushed ginger in any form smoothens your joints and muscles. It helps keep your muscles and joints relaxed naturally even after exercising.

However, this property does not mean it generally heals all sorts of joint or muscle soreness. It's can only be applied in cases that involve physical muscle struggles and therefore effects to incorporate it in the healing of other muscle and joint ailments may not be as successful because if the soreness is not an inflammatory then the ginger may not be of much help in that type of situation.

5.8.3 Treatment of Irritable Bowel Syndrome

Indigestion is caused by the lack of proper function of certain levels of the digestive system. It mostly causes some type of discomfort in the part just at the beginning of the tummy.

This type of pain is mainly caused by clogging of food that should have been digested processed and eliminating. These can be corrected by consuming some ginger which helps increase the rate of digestion because it is rich in fiber that is very essential in the digestion of food. Therefore, it solves the indigestion issue and also gives room for food. Also, when experiencing some difficulties eating then ginger may come in handy since it not only helps you digest food but also helps increase the appetite and boosts appetite. This will therefore help you with solving most of your food problems that you may be going through by just recommending yourself some ginger.

5.8.4 Treatment of Issues that are related to the Stomach

So basically, most stomach upsets are caused by small infections or bacteria. Ginger basically when ingested clears your tummy and therefore drains every type of infection that your stomach is trying to fight and therefore before even realizing it your stomach issues have been solved (Legwaila et al., 2011). Also, the stomach is a very sensitive organ that may seem to be disturbed by anything anytime. At times expectant women get nausea and morning sickness rapidly. You can also have nausea too expectedly and unexpectedly. Ginger, therefore, comes in real handy with its stomach remedies and helps treat any kind of upset in the stomach. The ginger also affects the nausea effect and neutralizes it with enzymes that help to get rid of the nausea feeling making the person more comfortable. It also helps reduce the cramps in females reduces the cramping pain that they go through during their periods. Traditionally, people used to use ginger tea as a remedy for all pains menstrual pains and it used to work for most of them since it helped calm them down.

5.9 LAVENDER

Briefly, *Lavendula angustiflia* is the most popular genus of *lavendula* and it is what we refer to as lavender. This plant is grown in northern Africa, Southern Europe and some sections of Asia. This plant is grown for its beauty and fragrance with even a color named specifically after it. Although it's a rare gem, it was used in traditional days as an herbicide for its medicinal value to the natives that grew the plant. Although currently there is barely any scientific evidence to back all the theories that have been established supporting the claim that lavender has healing properties.

Below we will look at some of the traditional medicinal uses of lavender although none of them has been conclusively scientifically tested and proven more research is still ongoing.

Figure 5.8. A representation of lavender and the health benefits.

Source: https://www.medicalnewstoday.com/articles/265922

5.9.1 Healing Wounds

In the tests carried out it was confirmed that actually, a wound that had the wound that had lavender on it sealed fast. This shows that the use of lavender on treating wounds can be argued since it facilitates the quick forming of platelets on the wound which therefore means it helps in the healing process of wounds that are open since it has a direct and eminent effect on platelets which are the blood cells that are responsible for the healing of an open wound (Malik et al., 2013).

5.9.2 Treatment of Anxiety as Well Insomnia

In most researches that have been concluded it has been concluded that actually, lavender has some tranquilizing ability in its scent that help people be calm and relaxed. Most relaxing spurs that teach people about peace and yoga have all used lavender. The lavender scent has some direct effect on your brain which helps you reduce the amount of anxiety going on through your mind and be able to take things a little bit more calmly than you would under normal circumstances. These tranquilizers in lavender help you to relax and therefore they also work perfectly as sleep enhancers since in a state of calmness your brain tends to more soundly than it would if it was in a restless situation.

5.9.3 Treatment of Various Infections

Although not a lot of research has been conducted on this some researchers do believe that lavender can come in handy in treating superficial skin infections, especially fungal infections that have proved to be untreatable by other fungal infection remedies mostly due to resistance. The theory is that lavender attacks the membrane of the bacteria making it impossible for it to survive and longer, therefore, ending up treating the infection organically.

5.10 BULB ONIONS

Referred to as *Allium cepa* scientifically, the onion is the most popular and rampant species of this classification. Although its origin is very unclear some people seem to agree on the fact that it originated in central Asia somewhere around Iran. Although it was originally analyzed and documented by Carl Linnaeus in the 1700s. This herb has been there for over 7000 years with mentions existing in some of the world's earliest manuscripts. In Egypt, for instance, it was used in their burial rituals representing the never-ending life even in death. What makes it even more fascinating is its mention in the Pompeii pone of the Roman empire's oldest manuscripts that signified continuance and also facilitated in their cooking(Mahapatra & Panda, 2012). The use of onions was very common considering that when the first Europeans landed in America, they reported that onions were there in plenty more than anything else. These show that that the use of onions ranges all over the universe.

Figure 5.9. A representation of the bulb onion.

Source: https://www.herbal-supplement-resource.com/onion-herb-benefits.html

Described below are some of the uses of onions in medicine as well as therapeutics.

5.10.1. Treatment of Issues that Are as a Result Of Inflammation

Onions help to reduce inflammation and therefore they come handy at relaxing the muscle. They help to calm the muscles and the nerves that control them. They keep the joints relaxed and even after physical fatigue the ingestion of onions can really facilitate the massaging and keeping the muscles feeling refreshed and also ensure that the muscles are not badly affected. It also helps in the treatment of asthma that affects the muscles of asthmatic people and, therefore, it helps them against having asthmatic attacks that are triggered by the muscles becoming inflamed.

The onions that you consume normally produce an enzymes that are antioxidants and therefore they can support and build the body's defense system so that it may be well equipped to face any kind of attack. Also, for people who experience a lot of allergic reactions, onions are a great remedy because they are a blue to produce mechanisms that can detect allergic reactions and prevent them from happening before they can affect you and it starts irritating (Mbuvi & Boon, 2009). These make onions a very immune-friendly enzyme that will boost your body's ability to face and every health issue that it faced by.

5.10.2 Treating Stomach Issues Such as Indigestion and Constipation

Like every other plant that is fiber endowed, onions come in handy in matters of the guts. Digestion is one long process that takes place from the mouth. Fibers are a very important part of this process and the lack of fibers leads to cogging in the guts leading to indigestion which some will system overload of the digestive system and all this will consequently lead to constipation and all these can be uncomfortable and the only way to decongest the system is by ingesting some fibers [onions in this case], which provide the required enzymes for the digestive tract to be washed clean and your indigestion issues are solved.

5.10.3 Treatment of Diabetes by Lowering Blood Sugars

Onion can lower the blood sugars whenever they shoot too high. Onions have enzymes that lower the amount of blood sugars in the body. If you test for the amount of glucose in the body after consuming onions, then it will be lower and this means that you do not have excess acids in your body and also your body sugars are regulated to low amounts which are more tenable by the body and therefore do not pose any challenge to the body (Mishra et al., 2018).

Chapter 6

HISTORY OF MEDICINAL AND AROMATIC PLANTS

CONTENTS

6.1 INTRODUCTION

Herbal medicine practices after ancient medicinal plants, also known as medicinal herbs, have been described and used. In protection against bacteria, fungi, insects and herbivorous animals, hundreds of chemical compounds are synthesized by plants. Many phytochemicals have been correlated with potentially or biologically confirmed behavior. However, because there is a wide range of phytochemicals in one single leaf, the effects of the medicinal application of the entire plant remain uncertain. In addition, the phytochemical content or pharmacological function of many medicinal plants remains untested in detail in clinical experiments to evaluate their effectiveness and safety (Prabha, & Kumar, 2015).

Common remedies have been used for thousands of years, food process and food storage, religious and cosmetic use, aromatic plants, herbs and essential oils. Plant extracts continue to be used as a source of medicinal compounds in food additives to fulfill the demands of western customers for natural, healthy and high-quality products. Research on the use and use of aromatic plants would also continue to clarify the possible welfare benefits for animals and humans. The Aromatic Plants tend to satisfy today's demand and future feed developments. Different terms, such as medicinal plants, spices, extracts or essential oils, are used by the international scientific community as a consequence of the omission or restrictive description or vocabulary of aromatic plants or herbs.

The earliest records of ancient herbs can be found in Sumerian history, which contains hundreds of medicinal plants, including opium, on clay tablets. Ancient Egyptian Ebers Papyrus, c. More than 850 herbal medicines are mentioned in 1550 BC. De materia medica, c., has documented over 1,000 recipes for medicinal products of more than 600 medicinal species served by the Greek physician Dioscorides of the Roman Army. 60 AD; this has been the basis of pharmacopeia for around 1500 years. Pharmacologically active drugs in the wild have been identified by ethnobotany in drug research and have been contained in hundreds of useful substances (Pandit & Acharya, 2008). They are popular prescription drugs such as aspirin, digoxin, quinine and morphine. Plants contain various kinds of chemicals, but most of them are found in four basic types of biochemistry: alkaloids, glycosides, polyphenols and terpenes.

Medicinal seeds are commonly used in non-industrialized crops, largely because they are easily available and cheaper than modern medicines. Thousands of plant species suspected of possessing medicinal properties are projected to export to USD 2.2 billion worldwide in 2012. In the year 2012. A total of $100 billion is expected in 2017 for global demand for botanical extracts and prescription drugs. Complementary drugs are not controlled in certain nations, but the World Health Organization is organizing a network to promote safe and just use. Medicinal plants face both general threats, such as climate change and the lack of housing, and the particular possibility of overpopulation to satisfy consumer demand.

6.2 PREHISTORIC TIMES

The herbal history of medicine, from the early days to the 19th century, is closely connected to the history of germ disease theory. Hospital treatment was based on evidence gathered by the experimental process from the 19th century to the present. Evidence-based use of pharmaceutical medicines, often from medicinal plants, has gradually replaced herbal therapies in western health care. But many people choose to use various kinds of herbal or synthetic drugs. The herbal component of these schemes is also significant. Because many herbs and spices historically used by humans for seasonal feed contain useful medicinal compounds, while the use of antimicrobial spices is an ancient response to foodborne diseases, the history of herbalism also overlaps with the history of food.

The tradition of an autonomous pharmacy dates back to the first third of the 19th century. Historically, the pharmacy had its beginnings in ancient times as part of medicine. The pharmacy's history fits perfectly with the medical history, but the two topics differ greatly. Pharmaceuticals is one of the most studied fields in the field of medicine, however, in contrast to its worldwide effects, its history on this specific subject is scarce (Saha & Sundriyal, 2012). Before the pharmacists came, the physicians consulted with medical priests and doctors.

Plants have not always been useful since ancient times, such as herbs and spices that are mostly used in cooking. Spices, especially in hot climates and more easily spoiled meat dishes, have been used in part to combat food spoiling bacteria. Angiosperms are the baseline of most herbal treatments (flowering plants). Humans are surrounded by herbal weeds such as nets and chickweed. Human beings were not alone as they wore herbs as medicines: certain animals, including non-human primates, monarch butterflies and sheep, consumed herbal plants when they were ill. Plant extracts from ancient gravesites are among the lines of evidence of herbal medicine in Paleolithic cultures. For example, Northern Iraq has provided large quantities of pollen from 8 plants, 7 of which are used as herbal medicines, at the "Shanidar IV" septic burial site. A fungus was found in Ötzi Iceman that had been frozen in the Ötztal Alps in his personal belongings for more than 5000 years. It is possibly against the whipworm that the mushrooms have been using.

Human experience is written before the use of plants as medicinal drugs. Archeological evidence indicates that humans used medicinal plants during the Paleolithic period, around 60,000 years ago. Plant recovered from ancient burial sites supported the claim that the Paleolithic people learned about herbal medicine (in comparison, other non-human primates are also known to drink medicinal plants to cure disease). In northern Iraq, for example, Shanidar IV developed large amounts of pollen from 8 vegetable plants, 7 of which are now used for herbal remedies, a Neanderthal funeral that lasted 60 000 years. More recently, Paul B Pettitt wrote that "A recent examination of the microfauna from the strata into which the grave was cut suggests that the pollen was deposited by the burrowing rodent *Meriones tersicus*, which is common in the Shanidar microfauna and whose burrowing activity can be observed today" These plants seem to have been used for the healing of intestinal parasites (Schumacher, 2011).

Prehistoric paleopharmacological findings promote the use of medicinal plants. In Shanidar Cave, for example, herbs like *Areca catechu* were found herbs and medicinal methods used for reflex, birds and beasts, and cool water, leaves, dirt and dust were used as a soothing agent.

Prehistoric medicine is the use of drugs previous to writing and medical history. The term 'prehistoric medicine' refers to a wide range of periods and dates owing to the time when writing by culture and location was invented.

Prehistoric medicine relies mainly on human pathology, fossils and remains. To gain insight into contemporary and historical cultures, anthropological studies have been carried out with previously uncontacted peoples and some Native communities living conventionally.

6.2.1 Disease and Mortality

There is evidence that many people have suffered from osteoarthritis, probably exacerbated by carrying large things that would have been a continuous and necessary work in their environment, and these diseases are now more common in prehistory than they are today. The transport of latte stones, for example, may contribute to the creation of micro spondylolysis in the spine, since this practice only started during the Neolithic period, which required a hyper-extension and a lower back torque when the stone was being moved. Things such as blues, slices and bone breaks can become very dangerous due to their insufficiency if they get sick with no antiseptics, appropriate equipment and germ sensitivity. There is also evidence that rickets, bone deformity and bone waste are caused by vitamin D deficiency (Osteomalacia).

Life expectancy was short in ancient times, about 25-40 years, and men lived longer than women; archaeological studies showed that many women and children died at birth and are thus blamed for a shorter life period than men (Shajib et al., 2013). Male humans, like hunters, have always obtained better food than females and thus may become less resistant to illness, so malnutrition may be a possible cause of lower life expectancy for prehistoric residents.

6.2.2 Treatments for Diseases

Plant material has been used to cure diseases in prehistoric culture (herbs and substances derived from natural sources). Historians do not completely understand which of the species used as plant material in prehistoric medicine rot quickly under most circumstances. The climate of the respective cultures can be studied and the species are already evolving under comparable conditions examined, and from the anthropological research of the existing indigenous groups, a speculative perspective can be obtained. Prehistoric populations, unlike ancient civilizations which may have developed plant material worldwide, may have been confined to regional areas, whereas nomadic tribes may have more plant material available than more stationary societies.

The effects of the various plant materials should have been observed as a test and error. Women who have taken care of the well-being of their families have collected and spread plant material in most cultures. Plants have been a major disease treatment throughout history. The Knowledge Fund has been distributed orally over decades.

Birch polypore fungus typical to alpine habitats in northern Europe may have been used as a laxative by ancient individuals, as brief diarrhea outbreaks are known to occur when consumed and among mummified males.

6.2.3 Treatment of Soil and Clay

Earth and clay may have given some of their original prescriptions to ancient civilizations. This synonym for geophagy is very common with both wild and domesticated animals. Geophagy is popular, particularly among today's non-human primates. They have learned of different treatment clays from early people to the study of animal behavior. Such clay is used internally and externally for wound care, for example during surgery. Geophagy and external use of clay continue to be common among indigenous peoples around the world and pre-industrial cultures (Srivastava et al., 2017).

6.2.4 Phase of Service

However, data indicate that a concentration of practice is taking place in Peru in ancient cultures around the world (sometimes trepanning). Several theories question the argument behind trepanning; it may have cured many disorders, including headaches and seizures. There are signs that the bone tissue is partially recurring in the operating hole and, at least occasionally, the survival of the operation has occurred.

Many ancient populations have been able to use clay materials to find fractured or injured bones where appropriate (geographically and technologically). Clay was replaced by a wounded area that was strengthened so that the bone could recover adequately without any intervention. In the Americas, pincers of certain insects were also used to shield wounds from infection; an ant was kept over it, where it was removed to hold the wound closed.

6.2.5 Sorcerers

The well-being of the tribe has been protected by medicine men (also witch doctors, shamans) by harvesting and administering medicines, performing little service, providing medical guidance and supernatural treatment such as charms, curses and amulets to stop evil spirits. Doctors begin a patient-based tradition with family members and family members in the community of Apache, as would probably have been the case in many others. It's made up of magic formulas, drums and prayers. The therapist further reveals how this disorder is being handled and how, for example, people are reminded of their personal history and possible abuses of their religion or tribal rules.

The tribe believed that they could touch ghosts or gods to use their supernatural powers to cure the patient, and eventually to remove evil spirits. If neither the procedure nor the trepanning was effective, the ghost was considered too heavy to be removed (Tao et al., 2009). Medical men could have been a central character in the tribal system because of their medical potential and, undoubtedly, because they could sense the spirits. They were necessarily transmitted orally for religious and medical education.

6.2.6 Dental Drug

Archeologists in Mehrgarh, Balochistan Province, have discovered in Pakistan today that the inhabitants of Civilisation Indus Valley have an early history in medicine and dentistry (c. 3300 BC). As he cleaned one of the masculine teeth, Prof. Andrea Cucina of the University of Missouri, a research anthropologist, made an observation. Signs of boiled teeth dated to 7,000 B.C.E can be found following studies in this field.

6.2.7 The Evidence of Difficulties

There is no documented evidence to discuss, by implication, the geological age of history. Historians are required to use other sources, such as human remains and anthropological findings of cultures occurring in similar conditions. There are a number of complexities in using the sources listed above.

Human remains from this time are rare, and many have certainly been destroyed or rendered worthless by burial rituals. There are no signs that prehistoric people, like ancient Egyptians, have mummified the dead for religious reasons and are preserved in pristine conditions; they are the most informative archeological facts. These bodies are able to provide scientists with (at the time of their death) subjects such as weight, cancer, height, diet, age, and bone conditions.

Prehistoric sketches, which use mineral paintings such as lime, clay and charcoal and feather paints, animal fur and twigs on cave walls, are not formally classified as 'presentations' and are assumed to have been for magical or theological reasons, have been revealed by others as 'written evidence,' while some have revealed some prehistoric medicine (thought to be a medicine man). Many of the cellar drawings of human hands reveal absent digits, suggesting that they have been cut off for sacrificial or utilitarian reasons, as in the case of Pygmies and Khoikhoi (Tiwari et al., 2018).

The writings of such peoples (like the Romans) can be used to explain medical activity in their present prehistoric cultures. People who live in a typical nomadic life have also been used as examples, but there are clear differences in the climate of nomadic people. Prehistoric people in Britain, for example, cannot be precisely equivalent to aboriginal people in Australia due to geographical variations.

6.3 ANCIENT TIMES

There are several hundred medicinal plants on clay tablets in ancient Sumeria, including myrrh and opium. There are over 800 herbal remedies in ancient Egyptian Papyrus Ebers, including aloe, morphine, bean beaver, knob, ginger, juniper and mandrake. Hundreds of pharmacologically active herbs and spices, such as curcuminoids in turmeric, have been used in Ayurvedic medicine from ancient times to the present, as recorded by Atharva Veda, Rig Veda and Sushruta Samhita. The Chinese pharmacopeia Shennong Ben Cao Jing reports herbal remedies such as cannabis chaulmoogra and ephedra. This has been expanded in the Tang Yaoxing Lun Dynasty. Aristotle's pupil, Theophrastus, wrote the first systematic text on botany in the fourth century BC, Historia Plantarum. Greek doctor Pedanius Dioscorides, who works for the Roman Army, recorded over 1000 recipes for medicines using more than 600 medicinal plants in around 60 AD in De materia medica. For more than 1500 years, the book remained the authoritative guide to herbalism in the 17th century.

6.3.1 Myanmar, Burma

The written Mesopotamian herbal research was carried out over 5000 years ago by the Sumerians, who produced clay tablets containing hundreds of medicinal plants (such as myrrh and opium).

6.3.2 Egypt

The troublesome language and translation of texts from this period and area are of particular interest to ancient Egyptian texts. These conclusions are founded on a lack of full knowledge of the Egyptian language: translations also include a basic approximation between Egyptian and contemporary meanings. Texts such as Papyrus Ebers appear to clarify and ease some of the theories regarding ancient medical practices, but there is little knowledge on physical history. Papyrus includes lists of diseases and their cures from limbo fever to skin disease and has information on more than 850 herbal remedies, including ginger, juniper, morphine, ricotta, aloe and mandrake. Treatments were mainly intended to relieve the patient from more common effects, and the condition itself was commonly believed (Viswanath et al., 2018). Details on the collection and preparation of certain therapies remain relatively unclear, as certain manuals available for translation indicate that the practitioner already has ample knowledge of how the operations are conducted and, thus, those methods do not need to be re-established. While reading the old text is based on modern Egyptian herbal knowledge, the Egyptian tradition is introduced to regions across the globe through trade and politics, which have affected and altered many medicinal cultures and have provided insight into the world of ancient Egyptian medicine. Herbs used by Egyptian healers were mostly local, though some were introduced by others, such as Lebanon. Apart from papyrus, evidence of herbal medicine is still used in tomb drawings and medicinal trace vases.

6.3.3 Indigenous People

Some herbs, like turmeric, may have been used as early as 4,000 BC in Ayurveda, India. One of the oldest texts available in Sanskrit, such as the Rig Veda and the Atharva Veda, describes the medical knowledge on which the Ayurvedic system was based. Indian herbalists including Charaka and Sushruta spoke of many other plants and minerals used in Ayurveda later in the 1st millennium BC. Sushruta Samhita, ascribed to Sushruta in the sixth century BC, is credited with 700 medicinal plants, 64 mineral preparations and 57 animal preparations.

6.3.4 The Chinese

In China, seeds that could be used for herbalism were discovered at excavation sites in the Bronze Age of China from the Shang Dynasty. The mythological Chinese emperor Shennong is believed to have written the first Chinese pharmacopeia, "Shennong Ben Cao Jing" 365 medicinal plants and their applications, including Epheda, hemp and chaulmogram in the case of Shennong Ben Cao Jing (one of the first effective treatments for leprosy). The Shennong Bencao Jing has been improved over the following decades, as was the case in Yaoxing Lun, the care of medicinal herbs in the 7th century.

6.3.5 Rome and the ANCIENT GREEKS

The Hippocratic Corpus serves as a collection of texts on the "father of western medicine," the Hippocrates of Kos. Hippocrates and his disciples, although both of them disputes the exact authorship of some of these texts, are representative of the fundamental principles laid down. There is little question that the common and commonly circulated therapies of the early Greek era are included in the recipes and remedies contained in the Corpus pages.

While certain herbs in Corpus are similar to those used for the curing of religious fields, there are significant variations in the lack of rituals, scriptures or chants used in remedies. This division of medical practice really represents the need of Hippocrates for justification and rationality (Yadav et al., 2017).

Corpus products include a multitude of spices, both originating in Greece and imported from exotic places such as Arabia. While certain imported products may have been too expensive for general household use, some of the preferred ingredients are more common and cheaper with St. John's Wort.

6.3.6 Gallen

Galen of Pergamon, a Greek doctor working in Rome, was undeniably prolific in his effort to justify his knowledge in all matters of medicine, and in his quest he wrote numerous herbal texts and their possessions, in particular his therapeutic works. In this text, Galen addresses the combination of both medical science and health protection with disease prevention. Although Galen's work and the four fundamental virtues in counseling address a broad spectrum of problems, pharmacists have been able to reliably measure the precise treatments and effects of their particular patients.

6.3.7 "Carystus dioculus"

The writings of the Diocles of Carystus were both long and plentiful. He was to be taken seriously with his guidance on herbalism and healing, with enough reputation to be referred to as the "the second Hippocrates" While the original texts no longer exist, over the years Diocles was frequently referenced by many medical scholars who brought us knowledge of his work in those parts. Diocles is thought to have first written the extensive herbal work often cited by contemporaries such as Galen, Celsus and Soranus.

6.3.8 Pline

Pliny the Elder's Natural History is a detailed guide to nature in one of the first texts of the Encyclopedia and also includes a wide selection of herbs that are useful for health care. Pliny's writings have a very broad base of knowledge with over 900 medicinal products and plants listed, enabling us to learn about ancient herbalism and medicinal practice. He called ill health "the best of all-natural operations" and opioid treatment "the state of peace and war between separate modes of life" "Pliny"

De Materia Medica, like Pliny, was developed by Pedanius Dioscorides, composed of more than 1,000 pharmaceutical, mineral and animal products. Durable treatments were used in ancient times and Dioscorides has been the leading opioid substance specialist for more than 1,600 years (Yadav et al., 2018).

For later century's herbalists and botanists Theophrasticus' History Plantarum, published in the 4th century BC, was the first systematization of the botanical world.

6.4 THE MIDDLE AGES

The Benedictine monasteries also preserved their medical know-how during the early Middle Ages in Europe, interpreting and copying ancient texts and increasing herbal greenhouses. Hildegard of Binge wrote about medicine for Causae ET Curae ('Causes and Cures'). Writers translated many classical Greek books, such as Dioscorides, into Arabic throughout the Islamic Golden Age and added their own commentary.

Although there are certainly medieval documents demonstrating the use of herbs, the precise reasons and meaning behind the production of herbal documents over the middle Ages have been controversial among scholars. The first is that the substance of these medieval texts has been translated directly from their classical counterparts without any study or analysis.

In comparison to texts such as the Herbarium with several plant chapters, the compilation of signs, habitat definitions and plant synonyms confirms the likelihood that herbals will be used in an educated manner. In the past, the Leechbook of Bald, Lacnunga, Peri didaxeon, Herbarium Apulia, Da taxone and Madicina de Quadrupedidus, and the Ex Herbis Femininis, Herbarius and Herbarius were widely used.

The Benedictine monasteries were an important source of medical knowledge during the early Middle Ages in Europe and England. Much of these monastic authors' efforts, however, focused not on giving specific new insight and practice, but on interpreting and copying ancient works of Greek-Roman and Arabic. Many Greek and Roman medical writings have been hand-copied in monasteries as other subjects (Yadav et al., 2018). As a result, the monasteries have proved to be medical centers and their grassy gardens have provided raw materials to help cope with simple problems. Around the same time, there was unregulated indoor and outdoor folk medicine, causing many travelers to settle for herbalists. They included "wise-women" and "wise men" who recommended herbal medicine along with beauty, delight, divination and counsel. Hildegard of Bingen was one of the most prominent women of herbalists. A 12th-century Benedictine nun, Causae ET Curae, wrote a medical text. During this period, herbalism was mainly practiced by women, particularly among Germanic tribes.

At that time, the Arab School, Anglo-Saxon Leechcraft and Salerno were open to three main sources of healing expertise. The great scholar of the Arabian School was Avicenna, who wrote The Canon of Medicine, the traditional reference work for medicine in the Arab world. The Canon is known for its systematic and biochemical research, the discovery of viral pathogens and sexually transmitted disorders, the advent of quarantine for the prevention of infectious diseases, experimental drugs, therapeutic trials and the concept of a syndrome for the diagnosis of specific diseases. Salerno has become a well-known medical and health school in Italy. Constantine the African was a student at the University, accredited to introduce Arab Medicine to Europe (Akinnifesi et al., 2002).

Factory research in the Middle Ages tended to concentrate on important observations. "In the 16th and 17th century an interest in botany revived in Europe and spread to America by way of European conquest and colonization." Philosopher became herbalists, and university professors studied plants of great value. Herbalists have begun the use of herbs for both medicinal and farm purposes. In the middle Ages, botanists were seen as herbalists, gathering, and growing, drying, storing and designing plants. Some have been experts in the diagnosis, identification and use of plants by morphology and their habitats. These texts, known as herbal materials, contained splendid illustrations and pictures of plants and uses.

The importance of plants for men at that period was underscored by botany and the practice of gardening; traditional herbal medicinal uses. During the Middle Ages, the growth of library culture spread across the medieval world. The method of translation is well known from its beginnings as a scholarly undertaking in Bagdad at the beginning of the 8th century and its propagation across the European Mediterranean bursaries in the 11th and 12th centuries. A collaborative effort is a location mechanism to be interpreted and added by a variety of entities. Yet how nature was seen in the Middle Ages remains a mystery.

6.5 EARLY MODERN

Many copies and compilations of textual and photographic manuscripts have been produced from various sources, both ancient and new. The translation is both a dynamic process and a technological practice that has made a major contribution to research in the Middle Ages; it has been the focus of frequent improvements and additions. They are known for their extensive medical knowledge in the Benedictine monasteries. The origins of current medical practices can be traced back to the monastic influence that has been helpful in the treatment of numerous human diseases that have grown in these gardens. They founded monastic schools and taught monks how to read Greek manuscripts in Latin.

Medieval botanical knowledge was closely related to medicine, and the plant was used mainly in remedies. Herbals have been classified by plant names, the definition of the attributes, herbal medicinal ingredients, protective properties and other components contained instructions for their preparation and use. A textbook has been developed for the effective use of herbal medicines. The medicinal invention of Dioscorides was a basic herbal product developed for practical use. Theophrastus has written more than 200 papers describing the features of more than 500 animals. He developed a classification system for plants with a focus on their morphology, such as their outline and purpose. Tomatoes, cinnamon, bananas, asparagus and cotton have been described in detail. Two of his most popular works, the Plant Survey and The Causes of Plants, have remained in Latin for decades. For several decades, Crateuas was the first to create a prescription book for medicinal plants, and his book influenced medicine. The Greek scientist, Pedanius Dioscorides described over 600 distinct plant species that were useful for herbal medicine, and his image was used for pharmacology and medicine as late as the Renaissance.

Monasteries have been developed as medical centers. These herbs were passed from nuns to monks, and their patients were well known and used. The average people had no interest in these charts; they were meant to be nuanced and for persons with herbal knowledge and personal experience. "The use of these herbals has been blamed for proving to be counterproductive, and several plants seem to treat the same illness, and 'the Western world doesn't like this feeling,' and these plants can have many benefits if they are used by skilled healers.

Another method of transmission was the transition of oral medicinal information between generations; it was used to transmit medicinal data from generation to generation. It is a common misconception to interpret early medieval medicine by naming texts (Ahmad et al., 2008). However, it is impossible to describe herbal explanations without prior experience.

The first pharmacy or pharmacy was founded in Bagdad during the Islamic Golden Age in 754 and was regulated by the state in the ninth century. The production of botany and chemistry in the Near East had greatly improved medieval Islamic pharmaceuticals. For example, Muhammad ibn Zakarīya Rāzi (Rhazes) (865-915) has worked to advance pharmaceutical research in the Middle East.

As a result of the scarcity of Greek medical texts and the rigid adherence to European practice during the fall of the Western Roman Empire in the 5th century, the medical conscience suffered during the Byzantine prohibitions in the region near Salerno, which in the 11th century was a popular hospital and medical school.

At the beginning of the 11th century, Salerno scholar Constantinos Africanus translated several Arabic texts which contributed to the transition from Hippocratic medicine to Galen's medicine. In medieval Europe, monks generally did not speak Greek, leaving only Latin texts, including Pliny's job, to the translation of Constantine. On the other side, Muslim Spain has made Arabic medicine well known.

The first type of written medicine was the Antidotarium; medicinal texts and styles were published in printing presses of the 15th century.

Pharmacy shops began to appear in Europe in the 12th century. In 1240, Emperor Frederick II issued a rule distinguishing the careers of the surgeon and the apothecary. Ancient pharmacies continue to exist in Dubrovnik, Croatia, within the Franciscan monastery that opened in 1317. The City Hall Pharmacy in Tallinn, Estonia, dates back to at least 1422 and is still the oldest continuous pharmacy in the world in its original premises.

The anti-pharmacology campaign has begun in Bruges, Belgium, where a new law has been adopted which prohibits physicians from preparing medicines for patients. The oldest pharmacy is said to have been founded in 1221 in Florence, Italy, in the Santa Maria Novel Convent, now home to a museum of fragrances. The ancient pharmacy of Esteve is now housed in Llívia, a Catalan enclave near Puigcerdà, a 15th-century museum that houses albarellos, books of ancient medicines and antique medicines from the 16th and 17th centuries (Altendorf, 2018). Florence is also born in a new reception area, the first of its kind to use the document as a guide for the care of patients in all pharmacies.

The Republic of Venice was the first nation to convey the presence of medicinal products to contemporary health policies. In fact, 13 secrets have already been sold to the Republic of Venice for sale.

From Grete Herball in 1526, the early modern era saw the rise of illustrated herbs in Europe. In 1597, on the basis of his popular The Herball or General History of Plants, John Gerard wrote an extended English physicist, published by Nicholas Culpeper. With early modern discoveries and subsequent trading in Colombia, many new plant medicinal goods have come to Europe, where horses, seeds and inventions have been handed over to the Old World and the Americas in the 15th and 16th centuries. The spices came from knob ginger and turmeric to America; coffee, tobacco, and coca went the other direction. The 16th century manuscript of Badianus applied to medicinal plants in Central America.

The Great Medicinal Era dates back to the 16th and 17th centuries, many of them in English and other languages, instead of Latin or Greco. Plants that were present in America were more commonly incorporated in the 18th and 19th centuries, as well as in Western medicine.

The first herbal to be written in English was Grete Herball from 1526. The two most popular herbs are The Herball, or General Plant History, by John Gerard (1597) and The Expanded English Physicist by Nicholas Culpeper (1653). In turn, Gerard's text was a pirated copy of a book by a Belgian herbalist, Dodoens, who took his illustrations from a German botanical work. There have been some bugs in the first edition due to the faulty arrangement of the two bits. His doctors ridiculed Culpeper with astrology, magic, and myths combining natural medicine, but his book had phenomenal successes, such as Gerard's and other herbs. In both the Discovery Period and Colombian Trade, modern medicinal plants have been introduced in Europe. The Badianus Manuscript, composed of Nahuatl and Latin in the 16th century, was an illustrious Mexican herbal text.

6.5.1 17th century

However, the preeminent role of plants as a source of medicinal effects steadily declined in the second century. It began with the Black Death, a death that proved unlikely to avoid the then dominant system of four elements. A century later, Paracelsus recommended effective chemical therapy (like arsenic, copper sulfate, iron, mercury, and sulfur).

6.5.2 The 1800's

Herbals have a lot of medical experience in the Americas, with few physicians and a lot between them. The books included Almanacs, the New Herbal of Dodoens, the new Edinburgh dispensary, Buchan's home algebra, and others. Indigenous Americans shared with colonizers, apart from European plant knowledge, much of this information was not published or collected until the 19th century. John Bartram was a botanist who studied compounds transmitted to Americans and also wrote about these plants in almanacs.

With the advent of a chemical study, the role of plants in medicine changed considerably in the 19th century. Alkaloids were isolated from a number of medicinal plants, with morphine poppy starting in 1806 and followed by ipekacuanha and strychnos in 1817. As chemistry advances, medicinal plants have discovered a variety of types of pharmacologically active compounds. Merck began the commercial production of distilled alkaloid medicines, such as morphine, in 1826. The synthesis of a compound first found in a medicinal plant began with salicylic acid in 1853. At the end of the 19th century, the vision of the pharmacy turned to medicinal plants as enzymes had altered the active ingredients by drying whole plants and the market for plant-purified alkaloids and glycosides had been produced (Banyal et al., 2019). Plant drug development continued to be important in the 20th and 21st centuries with large anti-cancer drugs from the Yew and peripheral cores in Madagascar.

The formalization of pharmacology led to a greater understanding of the underlying effects of medicines on the body in the 19th century. At the time, Samuel Thompson was a thoughtless, but thought-trained herbalist who had such an impact on professional activity that he renamed himself Thompsonians, both doctors and herbalists. They were distinguished from the "regular" physicians of that time who used calomelic bluttering and contributed to the short-term renewal of the scientific herbal medicine system.

6.6 19THAND 20THCENTURIES

Around five thousand years ago, the Sumerian clay sheet of Nagpur described the first written account of the use of medicinal plants for the preparation of medicines. It contained 12 formulations of drugs, including alkaloids such as opium, henbane and mandrake, corresponding to over 250 different plants.

A total of 365 drugs (dry bits of medicine), many of which are still used today, have been edited by the Chinese root and grass book 'Pen T'Sao,' published around 2500 BC by Emperor ShenNung, including rhei rhisoma, camphor, folio, podium, generous purple gentian, ginseng, jimson, and ephedra, etc.

The Indian Holy Books of Vedas deal with the care of plants in this area. Even today, different spicy herbs used in India are nutmeg, cinnamon, clove, and so on.

The Ebers Papyrus contains 700 plant species and medicinal goods, such as pomegranates, aloe, senna, garlic, onions, figs, wilves, coriander, junipers, centaurs, etc., written around 1550 BC.

Aromatic herbs, such as myrtle and incense, have been used in numerous ceremonies according to the Bible and the Ancient Jewish Book of Talmud.

6.7 HISTORICAL SOURCES RELEVANT FOR STUDY OF MEDICINAL PLANTS' USE

In the Homer Epics, the Iliad and the Odyssey, produced about 800 BC, 63 plant species are extracted from the Minoan, Mycenaean and Egyptian Assyrian Pharmacy. Any of them was named after the mythological characters of the epics, such as in honor of Elena, the backbone of the Trojan War, Elecampane (*Inula helenium* L. Asteraceae). Herodotus (500 BC) used for beaver milk, Orpheus for scent and garlic, and Pythagoras for sea onion, mustard and cob. Herodotus (500 BC) called for rice oil. As for the plants of the Artemisia family, which are known to be sturdy and well-being, their name is derived from the Greek term Artemis, meaning 'well.' The works of Hippocrates contain 300 medicinal plants described in physiological action (459-370 BC): wormwood and common centaur (*Centaurium umbellatum* Gilib).

Botanical science was founded by Theophrast (371-287 BC) in his books 'De Causis Plantarium' – Plant Etiology and 'De Historia Plantarium.' More than 500 known medicinal plants were classified in the books of the period. Among others, he described cinnamon, iris rhizome, false hellebore, mint, cardamom, pomegranate grass, scented hellebore, monkshood, etc. Theophrast stressed the importance of people using them by steadily increasing doses in order to illustrate the negative impact of herbs (Barua et al., 2019). He obtained the epithet of "the father of botany," given that he had considerable merit in the classification and description of medicinal plants because of his consideration of the subjects described above.

"Around 250 medicinal plants, such as aloe, snacks, flax, cocoa, pepper, cinnamon, gentian, cardamom, false hellebore, and more were mentioned by revered medical author Celsus (25 BC-50 AD)."

Dioscorides, the "father of pharmacognosist," was the most prominent herbal medicine scholar in ancient history, and he studied medicinal plants as a military physician and pharmacologist in the Nero Army, everywhere he went with the Roman Army, and wrote "de Matteria Medica" around AD 77.

'Pliny the Elder, a contemporary Dioscorides who passed through Spain and Germany, wrote 1,000 medicinal plants in his book 'Historia Naturalis.' At this time, the works of Pliny and Dioscorides incorporated all the knowledge of medicinal plants.

Galen (131 AD-200), the most distinguished Romanian scientist, assembled the first collection of effective medicines for equivalent or equal use (parallel drugs), De successanus (simultaneously a pharmacist). All of the replacements offered, in pharmacological terminology, are not complementary and utterly miserable. Galen had also introduced a lot of new herbal items.

The Slavs of the seventh century, *Samuz sativum* as cure and veratrum set, *Cucumis sativus*, *Urtica dioica*, *Achilea millefolium*, *Artemisia maritime* L, *Lavandula officinalis*, Sambuci were used in cosmetics such as rhyme, rosemary, moth, bees and spiders and *Aconitum napelis* for certain insect wounds. In addition, in the 7th century, the Rosemary of the Slavs was used in cosmetics as *Samuzia sativus*.

Ability in medicine, herbal production and pharmacy moving to medieval monasteries. Typical monasteries are based on 16 medicinal plants grown by medical doctors-mint, anise, savory, tansy... Care has been focused on 16 medicinal plants.

The founder of the renowned medical school of Salerno, Charles the Great (742 AD-814), ordered the planting of medicinal items in the state property of his 'Capitularies.' About 100 different plants, including sage, marsh onion, iris, mint, centaur, cowl, marshmallow, etc., have been described. The great emperor, especially liked the sage (L.)' (Salvare means "save, cure"). The sage, however, is a necessary herb in all the Catholic monasteries.

Arabs have also brought many new pharmacotherapy plants, many from India, a country in which they traded, but most of the plants have been of real medicinal value and have survived in pharmaceutical factories all over the world. Arabs used aloe, lethal henbane, cocoa, ginger, saffron, turmeric, potatoes, cinnamon, rheumatism, senna, etc. They used aloes, too. In comparison to purgative Helleborus flavor and euphorbium, for example, sennae folium has been used up to that point. Certain heavy-duty drugs have been replaced by items with mild effects (Bisht, 2020).

European middle-aged physicians, who mentioned more than 1,000 medical objects, consulted the Arabic works 'De Re Medica' by John Mesue (850 AD), 'Canon Medicinae' and 'Liber Magnae Collection is Simplicum ET Medicamentorum' by Ibn Baitar (1197-1248).

Macedonia has a great deal of pride in the work of St Clement and St Naum of Ohrid. They mentioned Nike's pharmacological code of 850 and shared his immense knowledge in the medicinal plant with his disciples and the masses.

Marco Polo's journeys to tropical Asia, China and Persia (1254-1324) and the discovery of America (1492), as well as Vasco de Gama's journeys to India (1498), introduced many medicinal plants to Europe. There have been botanical gardens in Europe, and attempts have been made to grow medicinal plants from old and new worlds, both domestically and through imports. A large range of new herbal products, Cinchona, Ipecacuanha, Cacao, Ratanhia, Lobelia, Jalapa, Podphylum, Senega, Vanilla, Mate, Tabacco, Red Pepper, etc., have been used for the exploration of America. The Quinine Bark of *Cinchona succirubra* Pavon was introduced for European health in Cortical China in the 17th century, when the Countess of Chinchon was the first to use this bark as Countess Powder (Behera et al., 2016). The quinine bark was quickly defeated England, France, and Germany, having many opponents of distinguished doctors—members of a variety of academies.

Paracelsus (1493-1541) was also a strong believer in astrology and one of the pioneers of chemically processed raw plant or mineral medicinal products. He continued to emphasize his faith in observance and also supported the concept of signature, the "Signatura doctrinae" called his mark, which implies their application to such diseases according to the belief of prescriptive medicinal drugs. For example, Haselwort is liver-like; *Hypericum perforatum* L is also ideal for liver disease. As the leaves of the plant John's Wort look like stings, it will be beneficial to treat cuts and stings.

The demand for medicinal plants in the middle Ages, especially between the XVIth and X8th Centuries, increased, while the elderly were mostly used as simple pharmacists, infusions, decoctions and macerations. Compound medicines also contained medicinal plants and medicinal products of animal and plant origin. The herbal substance Theriac has become expensive and dangerous to be sold by combining medicinal plants, animals and minerals that are endangered.

In his work Species Plantarum (1753) in the 18th century, Linnaeus (1707-1788) gave a short description and classification of the plants he described at the time. Species are described and named without any knowledge as to if any of them have been previously described. The first term used for naming the genus was polynomial, while the other polynomial term described the other properties of the fungus (e.g. the willow Clusius was named Salix pumila angustifolia antera). Linnaeus modified the naming system into a binomial. The names of each species contained, with a small initial letter, the name of the genus, the original letter and the name of the species (Boyd et al., 2007).

The knowledge and usage of medicinal plants was a turning point in the early 19th century. Further discovery, reinforcement and separation of poppy alkaloids (1806), ipecacuanha (1817), strychnos (1817), quinine (1820), granate and other plants were the basis of a modern drug store and then glycoside insulation. Many medicinal plant active compounds have also been detected through the application of chemical processes, including tannins, pulmonary and saponosides, etheric oils, vitamins and hormones.

Chapter 7

CULTIVATION OF MEDICINAL AND AROMATIC PLANTS

CONTENTS

7.1 INTRODUCTION

All objects in the entire globe are made up of different components. The collection of all components make up a system of the object. For the proper functioning of the object as a whole; each component ought to be functioning properly. This concept does also applies to living things. Each living thing, be it a plant or an animal; is made up of different subsystems; commonly known as organs; which function independently with their results being related in a way for the thriving of a living organism. In a situation where one component fails; the entire system is affected for not all components which make up a system are functioning properly.

In situations of such failures, something has to be done to fix the failed component to restore the entire system to its normal functioning. For instance, in living organisms, drugs are used in fixing the malfunctioning organ to ensure that the entire system of a living organism goes back to its normal functioning.

The atmosphere is made up of different components, which play a big role in the thriving of both living and non-living things in the globe. Such components ought to be balanced in such a way that they are in the appropriate percentages to provide perfect atmospheric conditions. However, in some situations, the components might end up not being balanced due to natural factors or unnatural factors.

In order to ensure that the components are balanced and that there exists a serene environment; the aroma ought to be perfect; in that, it is not too bad or too good, but it is balanced (Chakraborty et al., 2011). In situations where the atmosphere is contaminated and the aroma is polluted, there arises a need to avail components that will clean up the air and ensure that the aroma is good.

7.2 A BRIEF OVERVIEW OF MEDICINAL PLANTS

Medicinal plants, which can also be referred to as medicinal herbs, have been used for eternity in treating different ailments in all living organisms. As stated earlier, a living organism is made up of different components that all make up a system; in case one of the components fails, there is a need for it to be fixed to avoid further damage to the living organism.

Traditionally, there existed individuals who were gifted in identifying which plant is good for consumption by both human beings and animals; and what value would it offer to the living organisms. Not all plants were good for consumption, and not all plants good for consumption had medicinal value in them. Hence, the need for an herbalist who has that special gift of identifying which plant has medicinal value; which part of the plant had medicinal value; and what value it had.

The plants identified for medicinal use had more than one value. For instance, a medicinal plant could have been used in treating crops, animals or human beings.

During prehistoric times, the raw materials obtained from medicinal plants would be used either in their original form or by boiling them. However, with the advancement in the evolutionary ages, medicinal plants have ended being used in the manufacture of drugs eliminating the use of raw materials from medicinal plants in their original form.

In our modern age, medicinal plants are only known to specialists who have majored in the study of medicine. Since most of the drugs administered to either crops or animals or human beings are manufactured, the world has lost touch with medicinal plants.

However, most insects in farmlands; ailments in both plants and human beings; have gotten used to manufactured drugs to the extent that they have evolved and are no longer eliminated by such drugs (Challies & Murray, 2011). Such mutations have pushed many people to go back to the use of medicine from medicinal herbs in their raw form.

The few individuals in our societies who have retained the knowledge they obtained from their forefathers have played a great role in reviving the use of medicinal plants in their raw form in our current generation. Such revival has led to the conducting of more research on traditional medicinal plants. The results from the research activities have pushed many governments to opt for the growing of medicinal plants to ensure that there is high availability of herbs in our current society hence eliminating most diseases to crops, animals and human beings.

7.3 A BRIEF OVERVIEW ON AROMATIC PLANTS

The environment in which we live ought to have clean air, meaning the components which make up the atmosphere ought to be perfectly balanced. With clean air, there exist minimal respiratory ailments to both plants and animals.

In the prehistoric periods, people used to plant aromatic plants in their surroundings to ensure that they had clean air with a good aroma circulating in their environments. The need for fresh air went to the extent of some individuals planting aromatic plants inside their houses to ensure that, even during nighttime, they could inhale clean air with a good aroma.

With the advances in the evolutionary ages; individuals in our current generation have followed the path of manufacturing sprays using raw materials obtained from aromatic plants. However, the knowledge of aromatic plants is known only by a few in our modern age. This means that most aromatic plants have been neglected and the ones in existence are few in number.

With aromatic plants being few in number, most of the fresh air sprays which are being manufactured end up having only small percentages of raw materials obtained from aromatic plants, and high percentages being chemicals. Such sprays have contributed to the rising numbers of respiratory ailments since they contain dangerous chemicals that end up polluting the air instead of cleaning it up.

The effects caused by the modern sprays have been an eye-opener to our current society and pushed for the need to conduct research on aromatic plants so as to know their types, among other things. The results of most of the research conducted on aromatic plants have seen the rise in plantations that major in the cultivation of aromatic plants; so as to ensure that, most of the sprays manufactured are made up more of raw materials from aromatic plants and less chemicals (Crush & Frayne, 2011).

7.4 CULTIVATION OF MEDICINAL AND AROMATIC PLANTS

The increased demand for medicinal and aromatic plants has opened an opportunity to farmers, to not only major in the plantation of crops and rearing of domestic animals; but also to major in the plantation of both medicinal and aromatic plants.

Some of the major factors which have pushed for the cultivation of medicinal and aromatic plants, more so medicinal plants are; zero side effects from the use of medicinal plants; continued exploration of natural products to avoid further extinction of valuable natural products in the globe; and the fact that medicinal plants are easily available among the poor and are affordable to them as compared to manufactured drugs.

There exist several opportunities that have contributed to the cultivation of both medicinal plants and aromatic plants. The large numbers of people who are choosing herbal medicines over manufactured drugs have opened an opportunity for the cultivation of medicinal plants. The preference for natural aromatic products over chemical aromatic plants has also created a need for the cultivation of aromatic plants. Increased deforestation rates in the natural forests which are known to be home to indigenous medicinal and aromatic plants have created a need for the plantation of medicinal and aromatic plants so as to replace the ones which are slowly becoming extinct in most nations. The increased demand for raw materials from both medicinal and aromatic plants has pushed for their cultivation since there is an already available market with minimal supply (Chazovachii, 2012).

With the continued research activities, new and improved varieties of both medicinal and aromatic plants have been invented hence the need of cultivating them to ensure that they become dominant in the globe. With technological advancements, different agro-technologies have been invented which are believed to play a big role in promoting the growth of both medicinal and aromatic plants. Still, with agro-technologies, most farmers do not have to worry about not being able to plant other crops because such technologies provide them with ways of intercropping other crops with both medicinal and aromatic plants. The increased number of modern processing technologies serves as an encouragement to farmers for they are assured of processing the produce obtained from both medicinal and aromatic plants. The huge profit obtained from the growth of medicinal and aromatic plants is for sure enough motivation for any farmer to venture into the world of cultivation of such plants.

The advantages of cultivating medicinal and aromatic plants are many. One, with the cultivation of both medicinal and aromatic plants, there is the creation of employment opportunities in the farmlands, processing industries and transportation sector hence reducing the rate of unemployment in most nations. With the increased global demand for raw materials from both medicinal and aromatic plants, nations are assured of an increase in revenue which could be used for other development activities in the country. Since not all parts of either medicinal plants or aromatic plants are used, the rest could be used as food for domestic animals hence reducing the need of purchasing fodder. The other benefits obtained from the non-medicinal parts of medicinal plants could be used in reducing the cost of production increasing the profits obtained from the cultivation of medicinal and aromatic plants. The foreign exchange obtained from exports of medicinal and aromatic plants could be of great benefit to the exporting country. Cultivation of both medicinal and aromatic plants requires minimal labor force hence efficient workforce in such farmlands. Most products obtained from medicinal and aromatic plants have a long shelf life hence reducing the cost of preservations. Since the plants are used in curing diseases and eliminating pests, such farmlands have low incidences of pests and diseases saving on the cost of production. Most aromatic and medicinal plants are not that selective when it comes to the soil types, meaning that they can be grown in any type of soil and still manage to produce products of high quality. Parts used for medicinal or aromatic value cannot be easily stolen. For instance, the roots of a plant could be the one with medicinal value, hence the chances of a thief stealing them are minimal; for it would mean that the thief ought to uproot the entire plant. The taste of most medicinal and aromatic plants is not that good, hence the chances of the destruction of such farmlands by domestic animals are low (Chakraborty et al., 2011). Most of the technologies used in growing the new-found variety of both medicinal and aromatic plants are farmer-friendly and environment-friendly. Birds are known to be major destructors of farm produce, however, with medicinal and aromatic plants, the taste is not that good as stated earlier; hence the chances of such crops being damaged by birds are low.

7.5 IMPORTANCE OF CULTIVATING MEDICINAL AND AROMATIC PLANTS

The cultivation of medicinal and aromatic plants provides a sense of belonging in different societies. This is because it reminds them of the ways of their forefathers. Society does also enjoy the cultivation of medicinal and aromatic plants for it is easy, does not require incorporation of complex technologies or immense labor force. The cultivation of medicinal and aromatic plants has opened a source of livelihood for the members of society. It requires only basic information for both medicinal and aromatic plants opening a chance for the illiterate in our societies to practice it reducing the feeling of being left out in a society where most individuals are learned and almost everything is to be benefitted by the learned people.

With the cultivation of medicinal and aromatic plants, traditional knowledge of our forefathers is preserved. The marginalized regions of our societies are being given a chance to contribute towards the revenue collected by applying their deep knowledge in the growth of medicinal and aromatic plants. Through the growth of medicinal and aromatic plants, marginalized communities are also allowed to pass the knowledge they have on the identification of medicinal and aromatic plants, the proper ways of cultivating such crops; down to our current generation (Dagar, 2014). The conversion of traditional knowledge into a source of livelihood in our current generations ensures the preservation of the rapidly eroding cultural knowledge and practices which have played a big role in the globalization of communities.

Cultivation of medicinal plants and aromatic plants will ensure that herbal medicines are readily available to society and at a cheaper price. Also, as stated earlier, the chances of disease-causing microorganisms mutating due to the use of medicinal plants are low hence it will improve the health of animals, crops and human beings. As for the aromatic plants, the use of naturally manufactured aromatic products will improve human health since the products will be made purely from aromatic plants with minimal chemical components.

Traditionally, most herbalists were females. Cultivation of medicinal and aromatic plants will provide a platform for women to participate in an economy growing activity. This will also give them a voice when it comes to matters about the conservation of the biodiversity of medicinal and aromatic plants at the local and national levels. Women will be able to earn a living through the cultivation of medicinal and aromatic plants; for the labor required is minimal; women are known to be good in the selective collection, simple processing of products from medicinal and aromatic plants and the trading of such products. Women are also known to be good at conserving the environment hence their participation in the cultivation of medicinal and aromatic plants guarantees the production of quality produce.

The growth of medicinal and aromatic plants will fill in the gap that has always existed between women and men by ensuring that; both genders obtain the same information about the cultivation of the new varieties of both medicinal and aromatic plants; meaning they will be able to have same income reducing the over-dependency of women on men (Dadabhau et al., 2013). Apart from access to the same information; there will be equity among both genders in that; they will help each other throughout the entire process of production of medicinal and aromatic plants. For instance, as the men focus on harvesting, transportation of farm produce, the women can focus on sorting and the packaging of the products from medicinal and aromatic plants.

As stated earlier, the cultivation of both medicinal and aromatic plants will provide employment opportunities in most countries leading to a reduction of crime rates; an increase in the revenue collected annually; and improve infrastructure in different nations. With the cultivation of medicinal and aromatic plants, most organizations will have an investment platform whereby they can invest their resources in the production of medicinal and aromatic plants. This will improve their interaction with the marginalized groups in our societies hence educating them on other issues apart from the cultivation of medicinal and aromatic plants.

7.6 GENERAL CULTIVATION TECHNIQUES OF AROMATIC AND MEDICINAL PLANTS

The cultivation of both aromatic and medicinal plants needs to be conducted in specific ways that are not that complex. For instance, soil type ought to be determined for different types of either medicinal or aromatic plants that require different soil types. The method of cultivation does also vary in that; most of the newly invented plant varieties might require the incorporation of certain technologies to ensure that they thrive well. The climatic conditions are also a necessary factor that ought to be considered before the plantation of both medicinal and aromatic plants for different varieties thrive in different climatic conditions. As for the farmland preparations, both medicinal and aromatic plants require the farmland to be prepared in a specific way for the plants are mostly tough and fibrous.

Before cultivation, the seeds to be planted or the planting materials ought to be made available for the farmers early in advance in large quantities. This is because, some of the plants require their seeds or planting materials to be soaked for a certain duration before plantation (Davis et al., 2014). As for the large quantities, not all planted seeds thrive, so they are to be planted in large quantities to ensure that even though some might not survive the harvesting season, a large percentage of them survive. Still, before cultivation, farmers ought to be educated on the specific ways of taking care of their farm produce more so for the newly invented varieties of both medicinal and aromatic plants. Still, most of the technologies used for the production of medicinal and aromatic plants are not common to most farmers; hence the need to availing them and educating farmers on their use.

As for soil fertility, most medicinal and aromatic plants require soils with high fertility levels; and if there is a need for improving the soil fertility, organic means are encouraged to ensure that the products are organic with no artificial components in them. As for the use of organic ways of boosting soil fertility and farm produce, farmers should be educated about the preparation of compost manure and other organic components. The medicinal and aromatic plants selected for cultivation should meet certain requirements based on several factors. One is that, before the plantation of a certain plant, it should have been available in that area even if it is was in small quantities. Two, the plant to be grown should have multiple benefits in that, it can be used for medicinal purposes, aromatic purposes, fit for consumption by both human beings and animals, among other uses. Three, the plant should be evaluated based on its post-harvest value.

7.7 INDIVIDUAL CULTIVATION TECHNIQUES OF EACH MEDICINAL OR AROMATIC PLANT

For a better understanding of the cultivation methods for both medicinal and aromatic plants, we will discuss individual plants and the methods of cultivation applied in cultivating them; since each plant does have its own unique way of cultivation

7.7.1 The Brahmi Plant

The Brahmi Plant, scientifically known as the *Bacopa monnieri L, is* a medicinal plant whose raw materials have been used in the manufacture of drugs known to treat ailments such; mental disorders, epilepsy, anxiety neurosis and insanity (Das & Prakash, 2009). It is also popular because of its unique ability to boost human memory; with the leaves of the Brahmi plant being the ones with the ability to improve the receptive and retentive capacity of the human brain.

Figure 7.1: Image of the Brahmi plant.

Source: *https://www.woodvalefishandlilyfarm.com.au/product/brahmi/*

Apart from its use as a medicinal plant, it can also be categorized as an aromatic plant. This is because its flowers are used in the manufacture of hair oil and sprays which enhance the hair odor.

Brahmi plant has the advantage that it can be maintained as a perennial plant saving the farmer double the cost of having to plant every season.

The growth duration of the Brahmi plant is quite short. This is because the Brahmi plant is known to produce flowers three months after the plantation. and its produce is ready to harvest four months after the plantation season. The plant grows to a height of four to five centimeters.

Just like most plants around the globe, the Brahmi plant has different varieties. The most common varieties are the Subodhak and Prayashkti that are known as preferred by most people because of their high productivity and quality produce.

Brahmi plant, like most crops, is soil selective and thrives only under certain climatic conditions. As for the soil, the plant is known to do well she grown in poorly drained soil. It can also do well in waterlogged areas. The most favorable climatic conditions for the Brahmi plant are subtropical conditions which have: high temperatures ranging from thirty-three degrees Celsius to forty degrees Celsius; humidity ranging from sixty-five percent to eighty percent; and the plant does extremely well if its planted during the end of the summer season and the commencement of the rainy season (Dixon et al., 2001).

When it comes to land preparation, the land should be thoroughly plowed before the planting season and made free of any weeds whatsoever. Since the Brahmi plant is planted as tree cuttings, the land should be thoroughly irrigated a day before planting to make sure that the plant cuttings thrive from the word go.

The planting material of the Brahmi plant, which is obtained from the nursery; ought to be four to five centimeters long with a few leaves, nodes and roots. The planting material is obtained by cutting it from an older mother plant into small pieces together with its roots. As stated earlier, the planting area should be well irrigated before transplanting. Immediately after transplanting, the land should be irrigated through flood irrigation to ensure that the transplanted cuttings catch up. Since the cuttings are transplanted on the onset of the rainy seasons, the cuttings are allowed humble time to grow during the hot and humid season; and a month to the next rainy season, the product is ready to harvest.

Since the Brahmi plant is grown in regions with high temperatures, the plant ought to be irrigated immediately from the transplanting season close to the harvest season. If possible, the plant should be irrigated continuously, all seven days of the week.

Due to the continued irrigation done in farmlands growing Brahmi plants, weeds are ever-present. Farmers should weed their farmland after every five to twenty days.

The biomass of the Brahmi plant is high during the flowering season of the plant; hence the commencement of its harvest season in the late days of the flowering season, which os four months after the planting season. As for the produce, the stem part of the plant is part harvested.

The stem is ratooned in the upper portions around four to five centimeters high and the rest of the stem is left for purposes of regeneration; for as stated earlier, the Brahmi plant is a perennial plant (Fakir et al., 2018). However, if the harvest is done during the second time of production to the Brahmi plant, the entire plant is uprooted after harvesting.

The harvested stem is preserved by spreading it on the ground at room temperature for drying purposes. The stems could also be treated by exposing them to a temperature of approximately eight hundred degrees Celsius for a duration of around thirty minutes.

After treatment, they are further for around eight to ten days by being spread on a mat under a shade. After that, the dried materials are packaged into polythene bags depending on the size required for them to be sold at the market. The packages are later stored in cool and dry places with precautions to avoid infestation with fungi and insects.

7.7.2 The Tulsi Plant

The Tulsi Plant, which is scientifically known as the *Ocimum sanctum Linn* is a medicinal plant. It is used for the treatment of the common cold, tonsillitis, asthma, diabetes, and earaches. The oil from the leaves of the Tulsi plant is used in the manufacture of dental cream, mouth wash and toothpaste. The plant can also be used in the manufacture of insecticides, antibacterial drugs and manufacture of mosquito repellent creams.

Figure 7.2: Image of a Tulsi plant.

Source: *https://www.indiamart.com/proddetail/tulsi-plant-14488053162.html*

The Tulsi plant can also be categorized as an aromatic plant since its leaves, apart from having oil properties, they can also be used in the manufacture of perfumes.

Like most plants, the Tulsi plant has multiple varieties with the most common varieties being two; the Sri Tulsi and the Krishna Tulsi. The main difference between the two varieties is that the Sri Tulsi has green leaves whereas the Krishna Tulsi has purplish-blue leaves in color. Still, the Krishna Tulsi variety is mostly planted for commercial purposes with Sri Tulsi being planted mostly for domestic use. The Sri Tulsi is mostly referred to as the white type whereas the Krishna Tulsi is referred to as the black type.

The Tulsi plant can also be referred to as the Holy Basil, or Manjari or Brinda. The Tulsi plant is known to thrive in a variety of soil types. For instance, it can do well in rich loam soils; can also do well in poor laterite soil; can also do well in saline and alkaline soils which are moderately acidic (Galán, 2008). However, well-drained soils are much preferred for they are known to boost the vegetative growth of the Tulsi plant. Areas that are waterlogged are completely avoided by farmers for they do lead to root rot which results in retarded growth of the plant.

The plant thrives well in regions that experience fairly high rainfall and are humid. Research has it that, regions that have long days and quite high temperatures are quite favorable for the growth of the Tulsi plant. The high temperatures are said to boost the oil quantity produced from the leaves of the Tulsi plant. The plant has the advantage of being resistant to drought and frost, despite the fact that the most favorable regions for its growth ought to have high rainfall. Cool regions are said to be a disadvantage to the growth of the Tulsi plant in that they lead to the production of low quantities of leaf oil.

Tulsi plant is first planted in a nursery as seedlings before it is transplanted to the main farmland for its growth. The seeds ought to be fresh for they are known to deteriorate with time if they are to be obtained from the plant produce (Gera et al., 2013). By fresh seeds it means that; farmers are advised to purchase seed which has been treated and the factor of deterioration done away with- from the seed stores.

The nursery should have soil with high fertility, through the application of organic manure; since most medicinal and aromatic plants require the soil to be completely free from any artificial components. Since the seeds are minute in size, farmers are mostly advised to mix the seed with sand and sown at a depth of two centimeters above the ground. After the sowing of the seeds, additional organic manure is spread above the ground and the nursery ought to be regularly irrigated. The seeds take an approximately of eight to twelve days for them to germinate, and the seedlings are ready for transplant after six weeks.

The land on which the seedlings are to be transplanted ought to be prepared early in advance. The land should be of fine texture by the time the seedlings are being transplanted. Enough organic manure should be applied a week before transplantation and appropriately mixed with the soil to ensure that the soil is of high fertility. If possible, the land should be divided into small plots of convenient size.

As stated earlier, the seedlings are planted after six weeks. The seedlings ought to have at least four to five leaves before transplantation. Immediately after the transplantation phase, the plots should be irrigated. After the first irrigation, the seedlings are given some time to establish themselves after which a second irrigation is done. During the second irrigation, most seedlings will have survived, with only a few having dried up. Filling the gaps of the dried seedlings and the poor plants is done before the second irrigation.

After plantation, the Tulsi plant requires frequent replenishing of the soil. This is done by ensuring that the ground is covered in grass; more of an herbage. Farmyard manure ought to be frequently applied from the plantation period to the harvesting period. Before application of farmyard manure, farmers are advised to ensure that the compost is well decomposed before use (Gururaj et al., 2017). The first application of farmyard manure is applied weeks before the transplantation period; the second application phase is done few days after the first irrigation; and the last application phase is done few days after the second irrigation.

Apart from the main two irrigation phases, the plant ought to be irrigated depending on the moisture content of the soils and the plantation season-for in most cases, the weather patterns do fluctuate, and seasons expected to have high amounts of rainfall end up having low amounts of rainfall or no rainfall at all. If it is during the summer season, farmers are advised to irrigate their farmlands twice a week; and if it is during the winter season, farmers are advised to irrigate their farmlands once in two weeks. The grass cover, the mulch, is applied to conserve the soil moisture content and avoid drying of the plants, more so in the summer seasons. Few weeks to harvesting, farmers are advised to halt their irrigation routine to give humble time to the plant products to mature.

Weeds have always been a disadvantage to the farmer. Due to the continuous irrigation of the farmlands, the Tulsi plant is forced to share the ground with some weeds. To avoid the competing of soil nutrients and the sunlight with the weeds, farmers are advised to frequently eliminate the weeds in their farmland. At the nursery, if possible, there should exist no weeds; meaning that, the farmer should always watch out for weeds. After transplanting, most farms ought to be weeded after every two weeks. Application of mulch is of great importance for the mulch cover helps in suppressing the weeds from growing. Farmers are advised against the application of chemical herbicides in the elimination of weeds; for such herbicides end up lowering the quality of the plant and its produce. However, excessive weeding is somehow discouraged to avoid the drying up of the soil due to frequent exposure to sunlight during weeding. After the second irrigation, weeding can be put to an end because the Tulsi plant matures to be bushy thereby suppressing the weeds in its ways.

Tulsi plant is said to be quite delicate; hence maximum care should be employed during the harvesting season to avoid any form of contamination to the Tulsi plant. All surfaces that come into contact with the plant during harvesting ought to be cleaned thoroughly and frequently. The harvesting is done at the bloom stage of the plant to obtain maximum oil yield and quality oil produce. The first harvesting phase is done after ninety to ninety-five days after planting whereas the other harvesting phase is done after an interval of sixty to seventy days from the previous harvesting phase (Haque et al., 2018). The harvesting ought to be done on bright sunny days for the production of high quantity and good quality oil. Farmers are discouraged against harvesting if the previous day had rains.

The produce obtained from a Tulsi plantation is not that perishable for it can last for up to six hours in the field. However, more than six hours without any preservation method being put into place leads to degrading of the produce in terms of quality and quantity. The harvested oil ought to be distilled in a clean distillation unit that is free of rust or any kind of odor. Before steam distillation, the oil is decanted then filtered and later subjected to distillation. The distilled oil is subjected to treatment by treating it using either the anhydrous sodium sulfate or common salt. The treated oil should be packaged in bottles or cans that are made of either glass or stainless steel and should be stored in cool and dry places.

7.7.3 The Gudmar Plant

The Gudmar plant which is scientifically known as the *Gymnema sylvestre Retz* is a medicinal plant that is woody in nature and known to belong to the family of Asclepiadaceae. It known to have leaves that are simple, opposite, elliptic or ovate and hairy in nature. Its flowers are small and yellow in color with an umbellate cyme. The follicles of the flowers are terete, lanceolate and of approximately seven and a half centimeters long.

Figure: 7.3: Image of a Gudmar plant.

Source: *https://www.plantsguru.com/gymnema-sylvestre*

The leaves of Gudmar plant have been used as antidiabetics; treatment of: stomach aches, liver tonic, laxative, diuretic, as cardiac stimulants, jaundice and leprosy.

With increased demand of the produce of the Gudmar plant, the plant has become quite vulnerable due to destructive harvesting. This is one of the main reasons as to why most governments have opted for both large-scale and small-scale plantation of the Gudmar plant to ensure that most nations enjoy its numerous benefits obtained from its produce.

The Gudmar plant is known to thrive well in a variety of soil. The red sandy loam soils and the deep black soils are said to be the most ideal soil variety for the growth of the Gudmar plant. However, the plant is quite sensitive to waterlogged areas, hence regions whose soils encourage waterlogging are avoided. Its perfect climatic conditions are said to be found in the sub-tropical regions which receive rainfall of approximately six hundred millimeters (Hoang, 2008). The plant can also survive in dry areas however, under irrigation; meaning it is quite resistant to high temperatures. The perfect climatic conditions for the survival of the Gudmar plant are areas that experience medium to high rainfall that is well distributed.

Just like most plants, the Gudmar plant has different varieties. The main varieties have been categorized based on the leaf size of the plant. One of the varieties is the *Small leaves type*, whose leaves are approximated to be of a length of approximately one to three and a half centimeters and width of approximately one and a half to two and a half centimeters long. The small leaves type are known for their softness and are dominant in dry regions. The other variety is the *Broad and pubescent type* which has oval leaves of length ranging from three to six centimeters and a width of approximately three and a half centimeters to five centimeters. This variety is also known for its dark green colored leaves which are also pubescent, as the name suggests.

The Gudmar plant is first planted in the nursery before it is later transplanted to the main farmland. Mature seeds from the Gudmar plant are collected after the harvesting season and sown in small polythene bags or small plots. After the roots are mature enough, the seedlings are later transplanted into the main farmland.

Before transplantation, the land ought to be thoroughly plowed and leveled. The pits for the seedlings should be dug fifteen days before transplantation with a distance of around two to three centimeters in between the pit rows and one to two centimeters in a pit. The pits are filled with green leaves and topsoil and later mixed with well-decomposed organic manure. The pits are also irrigated a week before transplantation.

The transplantation is done during the onset of the rains. Since the Gudmar plant is a climber tree, it requires support for its growth and development, hence farmers have to support the plant using logs. To reduce the stress of having to support the climbers with logs, farmers are advised to plant the Gudmar plant between trees; more of intercropping.

After transplantation, the newly planted cuttings are immediately irrigated with additional organic manure applied on top of the ground so as to boost the soil fertility (Hossain et al., 2017). Depending on the amount of rainfall, irrigation should be regulated. For instance, if the rains are high, farmers are advised to water the plant after every five to six days; and if the rains are low, the rate of irrigation is increased to avoid drying up of the plant.

Due to continuous irrigation, farmers have to deal with weeds. Immediately after transplanting, weeding is done regularly until the plant has established itself. Once the plant is well established, weeding is done in quarterly intervals.

Gudmar plant does take time before it gets to the point of harvesting. In most cases, it takes up to two years for the Gudmar plant produce to be harvested. As stated earlier, leaves are the economic part of the plant. Once the plant is mature enough for harvesting, the leaves are harvested twice a year with an interval of six months in between the harvesting seasons. The first leaves harvest is done at the onset of the flowering phase of the plant. The second harvest is done after the flowering season. The leaves are harvested by hand picking method.

As for the processing phase, after harvest; the leaves are dried under a shade whereby there is sufficient circulation of cool and fresh air. The drying period lasts up to seven to eight days.

Gudmar plant can be cultivated for a period of ten to fifteen years under good management; still producing products of high quality and quantity.

7.7.4 The Lemon Grass

The lemon grass which is scientifically known as the *Cymbopogon schoenanthus L.Spreng* plant is purely an aromatic plant. By being an aromatic plant, lemon grass is mostly used in perfumery, manufacture of hair oil, scent and soap making.

Figure 7.4: Image of Lemon Grass.

Source: *https://www.medicalnewstoday.com/articles/321969*

There are three varieties of lemon grass. One of them is the *East Indian lemon grass,* scientifically known as ***C. flexuosus.*** The second variety is the *West Indian lemon grass*, scientifically known as ***C. citrates***. The last variety is the *Jammu lemon grass,* scientifically known as ***C. pendulus***. Apart from the three natural varieties of lemongrass, most nations, such as India, have invented their own variety by cross-breeding *C.pendulus* and the *C.flexuosus* to come up with a new variety that is known as ***C. khasianus***.

Lemon grass is known to thrive in regions with poor soils, which are mostly found in hill slopes. The soil pH is expected to range from four and a half acidity and seven and half basic. The grass can also grow as a vegetative cover over naked eroded slopes. The soil is not that much of a concern when it comes to lemon grass (Hilson, 2016). As for the climatic conditions, the grass thrives well in warm and humid areas that experience plenty of sunshine with rainfall amounts ranging from two hundred and fifty to two hundred and eighty millimeters per year. The rainfall is expected to be uniformly distributed.

The nursery bed ought to have fine soils and they should be a raised bed one. The fine soil should be mixed with compost manure together with leaf mold to ensure that the fertility of the nursery bed is high ensuring that the seedlings to be produced from the nursery are of high quality. The seeds are sown in lines with an interval of approximately ten centimeters apart. After sowing the seeds, the bed is further covered with cut grass materials.

The seedlings are ready for transplanting once they are about two months old and of a height of approximately twelve to fifteen centimeters. Before transplantation, the field ought to be cleared of any underground vegetation and pits of a depth of around five centimeters cube dug with a spacing of fifteen centimeters from one pit to another and nineteen centimeters from one pit row to another. The field soil should also be mixed with well-decomposed compost manure before transplantation. After transplantation, lemon grass does not require irrigation since it is naturally a drought-resistant crop.

Lemon grass is harvested after ninety days after planting. After the first harvest, the other harvesting phases are done with an interval of around fifty to fifty-five days. The grass cuttings are made at a height of approximately ten centimeters above the ground level. Lemon grass can be retained for up to five to six years if well managed and also depending on the soil type it is grown.

7.7.5 The Mint Plant

The mint plant, which is scientifically known as the *Mentha asiatica L* is an aromatic plant commonly known because of its uses as a flavoring ingredient to foods and drinks.

Figure 7.5: Image of mint Plant.

Source: *https://www.goodhousekeeping.com/home/gardening/a20705630/how-to-grow-mint/*

The plant does well in deep loam soils, although it can also thrive in sandy loam soils that are well-drained. The soils have to be well aerated and with a loose texture. Mint plant does well in regions that have temperate climatic conditions. This is because it requires plenty of rainfall during the growth period and good sunshine rates during the harvest season. Before plantation, the land is thoroughly plouwed till it has a soil of fine tilth. After land preparation, the mint plant suckers are planted with a spacing of approximately thirty centimeters from one plant to the other.

To keep the soil loose as required, frequent weeding is done to eliminate the competition between the weeds and the mint plant; and also to ensure that the soil is well aerated (Islam et al., 2019).

Mint plant does not require high rates of irrigation since its roots do not penetrate deep into the soil; so light and frequent irrigations are recommended.

Harvesting of the mint plant produce is done after three months since the period of plantation, and it is done two to three times a year. The interval between the harvesting phases ranges between eighty to ninety days from the previous harvest. The crop is cut at ten centimeters above the ground with the help of a sickle. Farmers are encouraged to harvest on bright sunny days so as to increase the quantity of oil to be harvested.

After harvest, the mint herbage should be dried before the distillation period; however, farmers ought to be alert to ensure that the drying period does not last long to the extent of decomposition of the herbage. The material left after the distillation of oil, which is referred to as spent grass can later be used as manure. The containers in which the oil is distilled ought to be stored in cool and dry places and away from light.

Chapter 8

REGULATION OF MEDICINAL AND AROMATIC PLANTS

CONTENTS

8.1 INTRODUCTION

Medicinal plants together with their extracts are considered to be very important in many developing countries especially when dealing with exports. In the current world, there has been a lot of changes with chemical drugs being present all over the globe. However, in the last decade to this date, individuals' preferences have shifted to natural health products and as such, many last developed countries are looking to enter into the trading or rather the exportation of what was once known as poor people's medicine (Jamnadass et al., 2011). The lack of awareness among the people in the world has led to the lack of markets for these products especially when dealing with international markets even though the demand for such products is quite high.

Medicinal and aromatic plants are claimed to be herbal drugs or rather medicine which are said to be ideally botanical raw materials. They are majorly used for therapeutic purposes, culinary purposes and other instances are components of medicinal products, natural health products, cosmetic products and medicinal products. Some of the value-added natural products are made by adding other ingredients and of course the medicinal or aromatic plants.

These value-added natural products are some of the products that are highly favored by women all over the world for the protection of their skins, their hair and generally their entire body. Some of the products include, natural oils, oleoresins and liquid extracts. For instance, the aloe vera gel constitutes 90% aloe vera liquids and some preservatives and oils. This shows clearly that there is an industrial demand for such products for the sake of meeting the ever-rising demand.

In addition to this, some of the traditional health care practitioners such as the witchdoctors use medicinal and aromatic plants in order to heal the members of the society that have gone to them for healing. Most people say, thank God for modern medicine. But what should be said is thank God for natural remedies. Most of these natural medicines are very useful in preventing some of the most feared kinds of diseases and also for curing them if properly studied with the researchers of medicine. Some of the finished products that are made from plant extracts are even sold over the counter with the trust that they will have no damage to the user of such products. However, it is advisable to anyone that would want to use any natural remedies to ask a doctor or rather a specialist for a prescription.

Figure 8.1: Aloe Vera gel is manufactured in an industry with the primary raw material being the Aloe Vera extracts from the Aloe plant.

Source: *Ayurcentralonline*

With this ever-rising demand for botanical products, the increase in demand for natural remedies is said to be skyrocketing and in years to come, the demand will be unimaginable. As such, some regulations have been put in place to take care of such plants and ensure that they are not misused by the public or rather the health workers. Some guidelines have been created for farmers and other collectors on various ways to produce high-quality botanical raw materials. These guidelines are known as Good Agricultural and Collection Practices (Kotwal et al., 2011). These guidelines aim to ensure that there will be no extinction of herbal remedies or rather the plants that are helpful to the entire humankind. With the growth in demand for herbal medicines over the last decades, the majority of individuals leaving in the rural communities have been presented with somewhat a golden opportunity for them to be able to generate income by cultivating and the collecting medicinal plants. With the growth of the herbal industry, some areas have drawn concerns among individuals that are using such products. The concerns have been on the quality of the raw materials that such industries use in the making of the herbal remedies. In most instances, there have been a number of cases whereby consumers of the products have been able to file lawsuits against certain companies for damage to their bodies after the use of such products. The claims were that the remedies that the individuals used were somehow contaminated or adulterated or the components that were claimed to have been used were not the ones. These cases are however not very popular but they may lead to the destruction of an industry's reputation or even the loss of customers. As such, the governments have been put in the spotlight to implement regulations for the herbal industries to pressure the industries into providing consumers with products that are safe and also with greater quality assurance. With the rise in demand also, the governments have been put in the spotlight to look at the effect of an increased collection of the products from the wild for commercial use. It is well known that when products are collected from the wild for commercial purposes, they are not collected in smaller quantities but rather in larger quantities. This however is not the problem. The problem, in this case, is that these industries collect these products but do not replace them. If possible, the industries should be obligated to cultivate these plants to avoid collecting them from the wild. This is so because, many valuable medicinal species are currently under threat, some are rare and even some of them have become endangered. This is because they have been heavily exploited in their natural habitats. Because of such concerns, the manufacture of medicinal products in many countries is highly regulated to protect such valuable species of flora in the world (Koohafkan & Altierim, 2011). The firms that are involved in the production of such products are currently required to adhere to very strict rules as instructed by their governments which is not making life easier for the farmers of such products. Currently, the suppliers of the raw materials are required to sell high-quality botanical raw material in order to be accepted by the firms. If the products are of low quality, then they are rejected. As such, the responsibility of producing high-quality botanical raw materials has fallen to the farmers and the primary processors of medicinal plants. The initial stages in the production of medicinal are plants are quite crucial in determining the quality of the end products that are produced by the herbal industries.

Figure 8.2: The herbal medicine industry is growing due to the rise in demand which is causing governments to get involved in order to protect the different plant species used to manufacture herbal remedies.

Source: *The Malaysian Reserve*

The main aim of the GACP guidelines is to ensure that all growers and collectors of medicinal products meet the tight demands of the industries and also ensure the protection of the plant species. The adherence to the GACP guidelines has been established to be voluntary. This means that some of the buyers in the herbal industry are strict to demand products that have high standards and some do not. I guess that covers the small percentage of products that individuals who use the products are complaining about. However, there are some buyers in the industry that are unaware of the standards that have been put in place by the government and therefore do not demand products that meet such standards. The best way to go around such instances is to ask different governments to have consultants that will be going around to different industries introducing the guidelines to them. Ideally, most of the buyers in the herbal industry are recognizing and adhering to the GACP guidelines by looking at them as a way of value addition to their products. This is so because most of the consumers of the final products could always demand products from industries that are known to follow the GACP guidelines. Due to the fact that these industries attract many customers, their demand for the raw materials becomes high which then becomes beneficial to the medicinal plants' growers and collectors as enables them to gain access to high valued premium markets.

Figure 8.3: The guidelines for different medicinal and aromatic plants need to be common across different variables so that farmers can produce the required varieties using the resources they have.

Source: *Krishi Jagran*

Medicinal and aromatic plants vary according to the place in which they are grown or rather collected. As such, the GACP guidelines tend to vary according to the different resources available especially when it comes to their implementation. The main objective of the GACP guidelines is to be able to present guidelines that can be able to apply to all species of plants that can be medicinal or have aromatic benefits, in all environments and to all producers of the same plants. Have different rules working for different categories could be very confusing especially to the producers. This even makes the producers and collectors of such products quit from such practices. With uniform guidelines, the farmers will then have the ability to apply the principles in their situations by use of the resources that are available at their disposal. With these facts, below are some of the correct and incorrect practices as recognized by the GACP guidelines. There are also simple and affordable ways in which these guidelines can be implemented.

8.2 GENERAL PRINCIPLES

Hygiene and Cleanliness are considered to be the most important principle in the GACP guidelines. This serves to be right because these are raw materials that are expected to be used on a person's body. Due to the sensitive nature of humans, the raw material needs to be clean and also kept in clean and also cool conditions to prevent the growth of any form of bacteria (Kalinganire et al., 2007). It is also very important that in every stage in which a medicinal plant passes, the person in charge at a particular stage maintains the hygiene. This is from the selection of seeds, the planting of the same, the maintenance, through to the manufacture of the same to get the final product. The GACP guidelines offer some form of very specific guidelines that are very useful in ensuring that the hygienic production of the medicinal product has been maintained. The main point that all the actors in the various stages need to remember is that, any equipment, or anybody that the medicinal plant material comes in contact with needs to be spotlessly clean. Anything that comes in contact with or the medicinal or aromatic plant is placed n needs to be thoroughly cleaned and sanitized before the entire process of manufacture. The equipment needs to be kept clean because, these plants will be used to make remedies useful in improving people's lives and should not be contaminated by any form of bacteria.

Microbes such as bacteria and fungi are quite difficult to be seen by the naked eye and therefore may exist on most surfaces. As such, if the medicinal plants come in contact with such surfaces, they too will be able to carry such microbes which will then be very harmful to use by human beings. To prevent such risks, buyers always check the bought plants by use of the microscope to ensure that they are not carrying any microbes. In case they find that the number of microbes present in a certain plant material exceeds a certain specified limit, then they are at liberty to reject the products from the supplier. Some of the dos and don'ts as stated by the GACP with regards to hygiene and cleanliness include:

- Maintaining personal cleanliness and hygiene while handling medicinal and aromatic plants, for example, clean body, clean clothes, bathing regularly, having well-groomed nails among others.
- Cleaning all surfaces that the medicinal and aromatic plant parts will come into contact with during and after harvesting. This also includes the equipment that will be used in the collection process
- Ensuring that the community in which the medicinal plants grow are maintained in such a way that they are kept clean.
- Before handling any medicinal plants, one should ensure that they wash their hands preferably with soap and water.

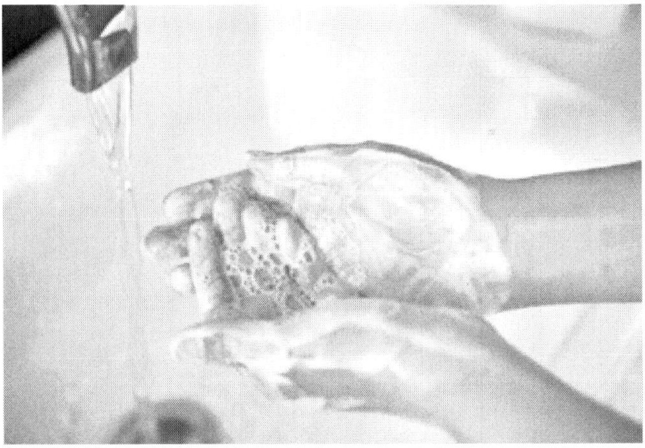

Figure 8.4: One should wash hands preferably with soap and running water before handling medicinal and aromatic plants.

Source: *CDC*

- All tools that are going to be used to handle the medicinal plants should be properly washed before they come into contact with any of the medicinal plants.

Some of the don’ts include:

- One should not handle medicinal and aromatic plants with dirty equipment
- One should not handle medicinal or aromatic plants with dirty hands
- One should not place the harvested medicinal and aromatic plant on a dirty surface
- One should not urinate, defecate, throw waste or even spit within the vicinity in which the medicinal and aromatic plants are grown. One should not allow any form of animal be it a cat, dog or even rats to urinate or come in contact with the medicinal or the aromatic plants
- In case one is suffering from an infectious disease or has an open wound then they should not be allowed to handle any medicinal or aromatic plants.

Correct identification is another crucial principle of the GACP. In other circumstances, one may find that there are different species of medicinal plants that are very identical to one another and even in other circumstances may even share a common name. The resulting effect is that; some other species may find their way into the manufacturing process that may then cause very serious damages to the consumer of such a product (Krishna et al., 2019). The only solution to such a problem is that, instead of identifying a plant species by the name that is known by the local community, it would be quite better if it is recognized by its botanical name. As such, it will ensure that the farmers and collectors are speaking of the same plant with the manufacturers. Some of the dos and don’ts in the GACP of correct identification include:

- One should send a sample to a government-recognized institute or a recognized NGO institution for verification if he/she is in doubt about the identification of a species. During this process, one can even learn from such institutes of the different varieties of such a plant that are of the same genus but are not medicinal.
- The seller should be able to ask their buyer to prepare a ‘plant monograph’ which has a description and photographs of the species that one will be required to grow or collect. This will help the farmer or the collector to know the species of medicinal plants that are demanded and also stop wasting their time growing or collecting other varieties.

- One should regularly visit a recognized government herbarium to make sure that they are collecting or cultivating the correct species of medicinal and aromatic plants.
- One should not be ashamed to seek guidance from an expert in the case where he/she has any doubt with regards to the identification of a species.

Some of the don'ts when it comes to correct identification include:

- If one is in doubt about the species of medicinal plant they are collecting or even cultivating, they should refrain from collecting it.
- In case you come across another species of plant that looks similar to a medicinal or aromatic plant or even shares the name, it is not recommended for one to use it.

8.3 GOOD AGRICULTURAL PRACTICES FOR MEDICINAL PLANTS

Majority of the medicinal and aromatic plants are collected from the wild. With the growing demand for herbal products, individuals are currently venturing in the production or rather the cultivation of such products because they are over-harvested and depleting in their natural habitats. GAPs for medicinal and aromatic plants tend to follow the same principles with regard to other plants. The only difference is that, for medicinal and aromatic plants, the plants are mainly grown for medicinal purposes and the parts need to be taken care of in such a way that they remain to be relevant and also safe to be used (Marín et al., 2008).

One principle of GAP is the prevention of contamination. Since these plants are used for the health of many individuals, they need care and protection from any form of contamination. In the process of cultivation, some risks are present that may lead to the contamination of the plant. As such, the farmer should be careful not to use pesticides or even place the harvested plant products in dirty containers. The second principle of GAP is the use of the best active ingredients. In the long run, medicinal plants will be used to manufacture herbal remedies. As such, during cultivation, the farmer should ensure that the plant parts that are properly harvested are the most relevant ones. Some of the factors that may influence the potency of the active ingredients include the variety of species used, the irrigation cycles, the cultivation site and the harvest time. As such, thorough research on the various factors that influence the potency of the active ingredients needs to be done.

Figure 8.5: Pesticides may cause contamination to the medicinal plants. one may decide to use natural remedies for pesticide management.

Source: *NDTV Food*

The third principle of GAP is that the farmer should be able to optimize on their yields and income. Some of the factors that have a significant impact on the yield include the quality of the seed used, the spacing between seeds, insect and pest management practices, irrigation practices, the harvest stage among other factors. The guidelines help guide the farmer to know what needs to be done at every stage of cultivation. Documentation and traceability is another principle of the GAP. This implies that the user of the final herbal product needs to be able to trace the materials used in processing back to their origin. This will make consumers conversant with the product they buy.

8.4 GOOD COLLECTION PRACTICES FOR MEDICINAL PLANTS

Some of the medicinal and aromatic plants are not recommended for cultivation. The case may be that these plants are difficult to propagate, take longer periods to grow, require a lot of input and therefore not profitable t the farmers (Pingali, 2007). With such instances, it is clear that medicinal plants will continue to be collected from the wild in years to come and as such the collectors need to be trained for better collection practices to protect the valuable resource for future generations. One of the principles of good collection practices is sustainability. This means that the collectors of medicinal plants should collect them in a manner that these plants may have the ability to regenerate. This will enable the plants to grow year after year. Another principle of GCP is the collection regulation. Some of the species of medicinal plants are endangered and some government regulations have been put in place to protect them. The collectors need to learn of such regulations. Other principles include the prevention of contamination, degradation and damage in which collectors are required to have clean harvesting equipment, the optimization of active ingredients in which the collectors need to consider certain environmental and climatic factors before collection and the documentation and traceability so that the users of the products can trace the products back to their origin.

Figure 8.6: During collection, medicinal plants should be collected in a way that they can regenerate especially when the collection is done in the wild.

Source: *Life Style-blogger*

8.5 GOOD POST-HARVEST PRACTICES FOR MEDICINAL AND AROMATIC PLANTS

The rising interest in the traditional systems of medicine and particularly herbal medicines in both the developed and the developing countries over the past two decades has bought concerns on the good agricultural and harvesting practices that should be carried out to ensure the safety and quality of a variety of medicinal plants and their products. This is essentially important because the sudden spurt in demand has led to medicinal plants being overexploited and many of them are been pushed to the brink of extinction (Sarker et al., 2017). While the quality of medicinal plants depends on the geographical origin, a cultivar of the species, stage of growth during collection time and post-harvest handling. Past studies have shown that hardly due attention is paid to the stage of maturity, processing and storage. The results are seen in the poor quality and unsafe medicines for mankind consumption as it will lead to more harm than good in their body.

As the collection of these medicinal plants from the wild areas continues to be a critical major source of medicine supply, their situation is very shaky, unsustainable and exploitative. Currently, about 90% collection of medicinal plants is from the wild. These plants are said to be generating about four million man days' employment (both part and full time) and since 70% of plant collection involves destructive harvesting, many plants have as a result become either endangered or vulnerable or threatened. The trade of these medicinal plant herbs is, however, vastly secretive, unregulated and expanding at a fast pace. Despite the concentration to this sector by humans as they supply a wealth of usable resources including biological, human and financial, the sector has not developed in the absence of sustainable harvesting practices, suitable processing and storage facilities and quality control of these medicinal plants (Sarker et al., 2017).

All medicinal plants should be harvested during an appropriate period so as to ensure the best possible quality of the medicine. While it is well known that the concentration of required chemical constituents in a medicine is strongly influenced by its developmental stage of growth as well as the season proper timing is very essential when dealing with these plants. The developmental stage of the plant directly influences the phytochemical concentration. For example, in *Andrographis paniculata* (Kalmegh) herbal plant which is used in several ayurvedic formulations as hepatoprotective herb, a study conducted by Pandey and Mandal in 2010 revealed that the maximum amount of andrographolide was found in *Andrographis paniculata* when harvested after 130- 150 days after planting. This time is during the time of initiation of flowering. Additionally, bacoside-A content of the herb was reported to be high from September through March and in June. Suitable harvest times for high yields of bacoside-A realized to be June and September to November. *Adhatoda vasica* ,which is a bronchodilator drug of Ayurveda shows wide seasonal variation in vasicine content in its leaves.

It exhibited higher levels of vasicine twice a year that is in March and in September. Interestingly, it coincided with the flowering time of the plant. This is because in March, it was full bloom condition and in September, it was partial flowering. During the vegetative stage, the plant is known to be containing a very low concentration of vasicine.

Figure 8.7: The medicinal plant needs to be harvested after 130- 150 days after planting which is the time of initiation of flowering.

Source: *Neutraceutical Business Review*

Besides, the stage of maturation of the plant parts to be collected is another necessary and important factor to be considered. The root of *Wihania somnifera* (Ashwagandha) herbal plant ,for example, is dug out just after 130-180 days after planting, while the stem of ,for example, *Tinospora cordifolia* (Giloe) is collected at full maturity, that is after 15 months. Past quantitative analysis revealed that all the phytoconstituents which are said to not be evenly distributed in the plant parts increased with the rise in the diameter of the stem except for the alkaloid content. Optimum production of metabolites never the less depends on the age and developmental stage of the medicinal plant. Harvesting of crude drugs with a higher concentration of active principle is a prerequisite in the preparation of efficacious medicines (Sarker et al., 2017). This ,therefore, calls for the medicinal plants need to be harvested at the period of growth or physiological developmental stage at which it accumulates the optimal amount of phytoconstituents that promises the highest medical results.

With the collection and harvesting being determined according to the quality and quantity of biological contents rather than the vegetative yield of the medicinal plant, proper care should be taken to ensure that no foreign matter, weeds or toxic plants are mixed with the harvested medicinal plant materials. The World Health Organization (WHO) have some of the following guidelines that need to be practiced during the post-harvest which entail firstly that any medicinal plants should be harvested by following sustainable harvesting practices. Secondly, these herbal plants should be harvested under the best possible conditions especially in avoiding dew, rain or exceptionally high humidity circumstances. However, in case of harvesting of these herbal plants occurs in wet conditions, the harvested material should be transported immediately to an indoor drying facility to expedite drying so as to prevent any possible deleterious effects due to increased moisture levels, which increases microbial fermentation and mold. Moreover, the collected/harvested herbal materials from the plant should not come into direct contact with the soils and in the case underground parts of the herbal plant for example the roots are harvested, any adhering soil should be scrapped off from the plants as soon as they are harvested/ collected. Contact with soil should be avoided as much as possible so as to minimize the chances of the microbial load in the harvested medicinal plant materials and where necessary large drop cloths that are preferably made of clean muslin are to be used as an interface between the harvested plants and the soil.

The Collected/harvested plant material should be placed in clean buckets/ basins, mesh bags, other well-aerated containers or cloths that are free from foreign matter which may entail even the plant remnants from previous collecting activities (Sarker et al., 2017). Immediately after the collection, the plant herbs may be subjected to appropriate necessary processing, including the elimination of undesirable materials and contaminants by handpicking, washing to remove excess soil, sorting and cutting the unwanted parts. The harvested raw medicinal plant materials should then be transported promptly in clean, dry conditions. Again, they may be placed in clean baskets, dry sacks, trailers, hoppers or other well-aerated containers and carried to a central point for transport to the processing facility. If the collection site is located at some distance from processing facilities, the necessity to air or sun-dry the raw medicinal plant materials before transport is highly recommended. If more than one part of the medicinal plant is to be collected, the different plant materials are to be packed separately and transported in different containers. Instances of cross-contamination should be avoided under all circumstances. Any mechanical damage or compacting of the raw medicinal plant materials as a result of ,for example, overfilling, stacking, loading of sacks or bags that may consequence in decomposing or otherwise diminish quality should be refrained.

Figure 8.8: Harvesting and handing cut flowers, should be placed in clean buckets/basins, mesh bags, other well-aerated containers or cloths that are free from foreign matter.

Source: *Mass Extension*

The decomposed medicinal plant materials nonetheless, should be identified and gotten rid of during harvest in order to avoid microbial contamination and loss of other good product quality. In addition, all the containers used for harvest /collecting these herbal plants should be kept clean and free from contamination by previously harvested medicinal plants and other foreign materials. If plastic containers are used, particular attention should be paid to any possible retention of moisture that could lead to the growth of mold. When containers are not in use, they should be kept in dry conditions, in an area that is protected from insects, rodents, birds and other pests, and inaccessible to livestock and domestic animals. The collected medicinal plant materials should be protected from insects, rodents, birds and other pests as well as from livestock and domestic animals. Collecting/ harvesting tools such as machetes, shears and mechanical tools, should be kept clean and maintained in proper condition. The parts that get into direct contact with the collected medicinal plant materials should be free from excess oil/grease and other contamination. Cutting devices, harvesters, and other machines should be kept clean and adjusted to reduce damage and contamination from soil and other materials (Sarker et al., 2017). These items should be stored in an uncontaminated, dry place or an area free from insects, rodents, birds and other pests, and inaccessible to other livestock and domestic animals.

For the case of the personnel handling medicinal plant material should they should firstly maintain a high degree of personal hygiene where importantly, they should be provided with suitable changing facilities and toilets with hand washing facilities to avoid contamination of the herbal plants. Secondly, these personnel members should not be allowed to work in the herbal material handling areas in the case that they are either suffering from or are carriers of a disease that is likely to be transmitted through medicinal plant materials such as diarrhea. Lastly, workers /personnel with open wounds, sores, and skin infections should not be allowed to be in contact with the herbal materials.

Figure 8.9: Personnel handling plants herbal parts.

Source: *Robert Harding*

Chapter 9

QUALITY, ADVERTISING, AND LABELING OF MEDICINAL AND AROMATIC PLANTS

CONTENTS

9.1 INTRODUCTION

Figure 9.1: Medicinal plants should be well labeled for identification.

Source: *https://www.wikiwand.com/en/Medicinal_plants*

Natural and fragrant plants are commonly used all over the world. However, these products have been accused of not having sufficient normative or logical evidence to approve their ingredients, safety and feasibility. A recent report found that 33% of the natural products tested did not contain the ingredients listed, and some were contaminated with artificial fillers. Manufacturers underestimate the significance of organic medicine labeling, while the scale of use is unknown by doctors. The chapter will study the complex nature of the aromatic and medicinal plant market, including the dangers and necessary upgrades needed to ensure the safety of products (Singh et al., 2014).

9.2 QUALITY CONTROL OF HERBAL MEDICINES

In general, the quality control strategies of natural medicines include sensory investigations (macro and microscopic evaluation) and scientific inspections using instrumental procedures, such as thin film chromatography, LC-MS, near-infrared (NIR), HPLC, GC-MS, and spectrophotometer.

Natural medicine and its derivatives have been used for many decades in Eastern countries, such as Korea, China, and Japan. Nevertheless, one of the key features of oriental natural medicine concoctions is that most herbal medicines are introduced into a compound formula in the form of a single spice or multiple spices during the preparation process, and they are separated with boiling water. This could be the primary reason why the quality management of natural medicines is more troublesome than Western medicines.

Although it has existed and continues to be used across the millennia, traditional medicine has not yet been endorsed in most countries. Therefore, education, training and research in this field have not received the necessary attention and support. The quality and magnitude of data on the safety and effectiveness of traditional medicines are far from adequate to meet the standards that support their use worldwide. Minimal research data is not only due to inadequate health care policies, but also a lack of appropriate or acceptable research methods for appraising traditional medicine.

When all is said and done, several markers or pharmacological dynamic changes or potential self-combinations are currently used to assess the quality and authenticity of natural medicines. Given that different ingredients are usually responsible for their beneficial effects, this quality can reflect the profile of herbal products. But, sometimes these different ingredients may "act synergistically" making it is almost impossible to separate the active components.

In addition, the flavor and potency of some natural products may vary according to the season of collection, plant origins, drying measures and other factors. Therefore, it is important to determine a large part of the phytochemical composition of herbal products to ensure their pharmacological reliability, and to comprehend their bioactivities and potential side effects (Tiwari, et al., 2018).

Along these lines of thinking, some chromatographic techniques (for example, high-performance liquid chromatography (HPLC), thin-layer chromatography, gas chromatography (GC), and capillary electrophoresis (CE) can be used for such documentation. According to these lines of thinking, completely natural products can be regarded as dynamic "compounds".

9.3 CHROMATOGRAPHY AND COMPOUND FINGERPRINTS OF HERBAL PRODUCTS

Since a single herbal medicine may contain a large number of natural ingredients, and the combination of different flavors may promote the mixing of hundreds of ingredients during the concentrate placement process, the fingerprints provided by the chromatography instrument may introduce a generally fair portrayal of the substances in medicinal plants.

9.3.1 Micro Thin Layer Chromatography (TLC)

Before establishing instrumental chromatographic strategies such as GC and HPLC, it is important to determine the conventional techniques for herbal medicine analysis. Indeed, even in the last few years, due to various pharmacopeias, such as the American Herbal Pharmacopoeia (AHP), it is still necessary to always pay attention to checking natural plants and still use herbal medicine trademarks. TLC has a lot of potential for the decomposition of herbal products. Moreover, TLC is quite basic and can be used for various sample tests. For every plate, around 30 spots can be concentrated at one time. Along these lines, it is well-known to use TLC to study natural prescriptions. Through technologies like CAMAG video store framework, and the TLCQA-UV technique, useful subjective and quantitative data can be obtained from the plant.

In addition, with the help of image analyses and digitization techniques created in software engineering, the similarities between medicinal plants can also be assessed. The benefits of using TLC include its simplicity, flexibility, high speed, sensitivity and modest ingredient preparations. Therefore, TLC is an advantageous strategy to determine the quality of natural products and conceivable deterioration.

9.3.2 Optical Equivalence Technology

The model of phytoequivalence was created in Germany to ensure the consistency of natural products. As this concept points out, chemical profiles should be established for natural products, such as chromatographic unique markings, and analyzed by clinically proven reference product profiles.

In 2004, the Chinese Food & Drug Administration (SFDA) announced that they will use strict quality standards (such as substance testing and standardization) to control herbal medicine (HM) fixatives. For this reason, HM liquid injections should be done forcibly. Similarly, in different testing strategies, it is strongly recommended to use chromatography technology to discover the fingerprints of plant medicines. The most effective chromatographic method for herbal medicine (HM) is the chromatographic example of plant parts through pharmacological activity.

Similarly, in different research procedures, it is strongly recommended to use chromatographic strategies to find fingerprints of natural substances. By definition, the chromatographic unique label of herbal medicine (HM) is actually a chromatographic example of a concentrate of certain compound fractions of pharmacokinetic or potential synthetic properties (Hossain et al.,

2017). The chromatogram should be highlighted with the basic attributes of "integrity", "fluffiness" or "similarity" and "contrast" in order to represent the HM examined.

It is recommended that in the case of obtaining Chromatographic Fingerprints, regardless of the total sum or potential convergence of the artificial trademark components and the various constituents of the herbal medicine (HM) (for example the "fluffiness"), the Chromatographic Fingerprints should be "equivalent". It can accurately guide the Verification and Identification of plant medicine ("legitimacy").

Therefore, it is significant to consider the different components in HM extracts, instead of considering only a few markers individually to determine the overall quality of herbal medicine products. Nevertheless, in any HM and its extracts, there are hundreds of unknown components and many of them are low in content. In addition, there is often variability in the same herbal material. Therefore, it is not an easy or trivial task to obtain reliable chromatographic fingerprints representing pharmacological activity and chemical characteristic components. Luckily, chromatography has quite powerful separation capabilities, so the complex chemical compounds in the HM extract can be divided into various basic subcomponents.

In addition, in recent years, combined chromatography and spectroscopy methods have been applied, such as gas chromatography-mass spectrometry (GC-MS), capillary electrophoresis-diode array detection (CE-DAD), and high-performance liquid chromatography-diode array detection (HPLC-DAD) which could provide additional spectral information, which is very helpful for qualitative analysis. In terms of eliminating spectral interference, selectivity, retention time-shift correction, chromatographic separation ability and measuring accuracy, the process has shown significant improvements with the help of spectral measurement.

9.4 MARKETING AND ADVERTISEMENT OF MEDICINAL PLANTS

The medicinal plant market has always been very large globally. But because a large number of small and medium-sized organizations in the business are unwilling to share information, it is often difficult to investigate data related to the herbal plant market. Wholesalers are also reluctant to provide data in the thinking that organizations may avoid them. In addition, cross exchanges between organizations are usually required, which increases the trouble of understanding the business. In general, producers know little about the actual sources of their therapeutic plants. Most producers in North America and Europe buy from wholesalers, the largest of which are from Germany. The world's largest markets for medicinal plants include France, Germany, China, Italy, Spain, the United Kingdom, Japan, and the United States. In particular, Japan has the highest per capita utilization of botanicals in the world. In the United States and Europe, given the universality of botanical treatments, the trade usually develops at a rate of 10% per year (Legwaila et al., 2011). However, due to the increasing official recognition of the advantages of natural treatments, their popularity may grow even further.

It is conservatively estimated that as of 2000, the currency value of global trade-related to medicinal and aromatic plants exceeded 60 billion U.S. dollars. As people's demand for medicinal plants continues to increase, whether in South Asia or internationally, this trade is expected to increase to 5 trillion by 2050. The Chinese market is very large and is owned by public and private owners. 13 of the top firms making Traditional Chinese medicine (TCM) are publicly listed on the local stock exchange, while 14 are government-owned. The total output of cultivated and wild medicinal plants in China is 1.6 million tons per year.

9.4.1 Strategies for Successful Marketing

- Medicinal Aromatic Plants (MAP) Supply Chain

  Various MAP species have unique prerequisites for the development, resource management, harvesting, processing and supply chain operations. In regional MAP markets, the distribution channel is usually short and the people involved are similar (small-scale farmers). To enable various people, gatherings or associations to complete unique exercises, and to communicate fairly, the contributions (labor, professional input, marketing skills, etc.) should be reasonable. Examples of stakeholders that may be included in the supply chain of MAP include; • Individual or collective farmers, harvesters or processors. • Private or network owners of plant medicine resources, and • Communal distributors, who acquire and transport products from the market, or external traders who may develop new trading routes.

  In rural, suburban and metropolitan environments, most medicinal plants are exchanged in small amounts within the region. For instance, it is estimated that the MAP supply chain in Southern

Africa is controlled by 500,000 traditional healers who distribute herbal formulas from more than 1,000 crops to around 100 million consumers (Malik et al., 2013). Therefore, small-scale farmers or potential traders wanting to enter the business often encounter various obstacles such as lack of regulation, substandard products and market saturation.

Similarly, export-oriented advertising has special needs, requiring detailed information about specific markets, product specifications and principles. A key test for those doing the MAP business is to identify these obstacles and, where possible and reasonable, find ways to eliminate them. Supply chain inspections can improve the understanding of how trade networks operate, the identities of major players and influencers, their specific activities, and various transaction processes. Special guidelines may be created, and the skills, limits, and experience needed to successfully participate in the MAP trade are essential.

9.5 CASE STUDY

9.5.1 The supply Chain of Vinca Rosea (*Catharanthus roseus*)

Catharanthus roseus, also known as Vinca Rosea, is a popular flowering plant in India, which is found almost everywhere. Although it mainly grows in Bihar region, few people actually recognize its medicinal value. Studies have shown that this plant contains about 65 alkaloids, among which indole, robinsine and serpentine are more potent. The leaves of VincaRosa contain alkaloids, such as vincristine and vinblastine, which can be used to treat some forms of cancer. In the Patna district of Bihar, the main cash crop is sugarcane. Nevertheless, due to falling prices, sugarcane production in Patna and surrounding areas is declining.

In order to help farmers get rid of this endless cycle, experts proposed the adoption of Vinca Rosa, which is economically feasible. This was a way to help farmers improve their status by adopting higher-value commodities. A total of three criteria were used to evaluate the plant products with the best possible returns. First, there must be market interest for the crop. Second, the harvest should be suitable for the current agro-climatic conditions (Mahapatra & Panda, 2012). Third, the production system used should generally reduce the risk to farmers with minimal effort.

The methodology adopted was as follows: I) To assess and subsequently develop a financially viable network for the production and collection of *Catharanthus roseus*, and other therapeutic and sweet plants through farmer associations, II) Organizing producers to form farmer interest groups (FIGs) III) Assessing the market demand for specific medicinal and aromatic crops; IV) Assessing the potential of producing and trading Vinca Rosa in the region V) Training FIG members to produce and manage Vinca Rosa; VI) Monitor production and post-harvest processing into final products; VII) Evaluate the potential to increase the amount of farmers enrolled in the program through exposure visits and promote expansion accordingly.

When the project was launched in 2000, about 60 farmers formed a federation, and began the development of Vinca Rosa throughout the Patna region. Within a year, 10 more groups joined this new program which aroused greater interest. By April 2003, ATMA had formed a network of roughly 50 farming groups. As the development encouraged more farmers, the overall number of new members signed to the project increased rapidly. Currently, there is a farmers' association advancing the development of medicinal and aromatic plants as a way to grow farm income and upgrade rural communities.

Resource management plans

In order to harvest wild MAR resources, the availability of wild species and their recovery rate after harvest should be used to offset market demand. Where business harvests are proposed or previously conducted, management plans are required to accurately assess the impact and establish a demand-sustainable harvest rate (Mbuvi & Boon, 2009). Effective implementation of a large number of components of the board plan will require specialized expansion support. This general practical guide can be used to develop a management plan, specifying how to estimate crop populations, distribution, and harvest yields.

In general, the root of wild ginseng is five to several times more valuable compared to propagated roots, and the explanation is social in nature. The Chinese people are the largest consumers of wild ginseng. They believe that the comparability of rough wild roots with human bodies represents the strength of roots. Cultivated roots cannot meet the trademark requirements of wild roots, and therefore cannot be deeply coveted by consumers. In Botswana, for example, traditional medicinal experts announced that the materials developed were unacceptable because the capabilities of this plant are not comparable to those of plants extracted from nature.

Harvesting Techniques

Figure 9.2: Rural collectors harvesting medicinal plants in the wild.

Source: *https://mountain.org/medicinal-aromatic-plants-program/*

The nature of the MAPs and extraction sustainability are closely related to the way plants are harvested. Most MAPS can be collected without damaging or destroying other plants, but it usually depends on which parts are collected to understand the potential impact on a single plant. For instance, harvesting a single plant to extract its properties may adversely affect the ecosystem's biodiversity since animals and plants are interrelated.

Moreover, the bark is habitually removed to extract its restorative properties, and intense and continuous harvesting often results in the degradation of trees, which may degrade from the constant scraping. In addition, the time of harvest and other processing aspects such as the speed and temperature of drying and storage conditions also affect the quantity and nature of the yield.

The season, harvest period and the techniques used vary, based on whether the tender parts are developed or harvested, and some restorative herbs are obviously collected in the evening. Farmers should collect roots regularly by digging 15 to 20 cm away from the plant (Mishra et al., 2018). Besides, exudates such as resins can be transferred to plastic containers or reusable metal tanks using few containers. In many cases, the harvested items must be dried for a short duration to be suitable. Some useful harvesting tips for medicinal plants include:

- Doing a stock inventory of the population abundance and distribution.
- Assessing tenure, risks and access based on available knowledge;
- Design and provide management plans, including: yearly harvest quotas, regional and seasonal restrictions, and domestication for certain plant parts or sizes;
- Implementing adaptive management and performing recurrent monitoring and evaluation.

Processing and Packaging

Using a range of technologies and equipment from basic to very specialized, processing MAP can add value, stabilize qualities to extend shelf life and improve hygiene. Production facilities may range from simple equipment for extracting essential oils to highly complex factories for the production of perfumes and medicinal preparations. Local treatments may involve preparation methods, such as hot and cold water extraction, crushing plants and squeezing the juice, pulverizing the dried plant material with water, oil or honey, and using a sugar source to turn the dry powder into a paste, and even plant fermentation (Prabha, & Kumar, 2015).

Therefore, making doses from extracts may be a suitable practice for small businesses, thereby helping to meet the medical and healthcare needs of local residents. Since medicinal compounds are usually naturally present in plants in small amounts, processing may require a large amount of raw materials, so expanding the scale or expanding the scope of activities may have hidden risks to natural resources and infrastructure.

Preparation and Processing of Medicinal Aromatic Plants

Advanced processing of medicinal plants can be compressing into pills or capsules to extend their shelf life, and provide better dosage effects. Then they must be stored in a tamper-proof container; • Suspension – a mixture of powder and water, including some type of preservative, does not need any special equipment, plus the mixture can be done directly before use;

- Creams and gels -The mixed or liquefied plant part is combined with an aqueous cream or gel. It can then be packaged in a plastic tube or container for easy application to the skin, where it is easy to absorb the medicinal properties;

- Tea – a simple processing process that makes the leaves dry, ready to be added to boiling water. Putting loose tea leaves in the bag can ensure a uniform dosage, but the equipment needed is dedicated and expensive, so it is best to leave it for products that are traded in huge quantities
- Tincture, syrup and fatty oil-Tincture is an alcoholic herbal extract, and syrup is a tincture with lower alcohol content. Like petroleum extraction, they are generated through extraction, fractionation, cold processing, purification, and other physical or biological procedures, plus are preserved with alcohol (Pandit & Acharya, 2008). In the simplest form of processing, they may include preparations done through soaking or heating herbal materials in alcoholic beverages and/or honey or other materials, which are stable and easy to manage, and are liquid to make dosage adjustment simple. It is best to pack them in brown glass bottles for protection against direct light, and it is ideal to use the dropper top to help apportion small amounts.

Drying

Drying is the main preparation method of medicinal and aromatic plants, since many specialized processing innovations require dry MAP materials. Some of the options available include cool drying, oven drying and freeze-drying. Oven drying is quicker and more effective compared to shade drying, and therefore includes less fungal contamination or the risk of "aflatoxin" infection. Nevertheless, large drying ovens are for small manufacturers/processors. Freeze-drying is the most expensive and most comprehensive drying technology. It can reduce plant materials to extremely low moisture content and provide stable materials with reduced risk of infection and strong medicinal properties, plus the plants can be used for a long time. In the original natural structure, or in a basic processed structure (diced, ground or boiled extracts).

Collectors usually sell plants to traditional healers knowing that these healers must take appropriate steps to produce medicines that are fit for administration. The packaging is typically simple, using locally accessible free products like newspaper wrappers, plastic packaging and recycled glass bottles. In contrast, medicines produced by commercial manufacturers must strictly comply with the regulations on processing and packaging.

Branding and labeling

Not many MAP merchants or therapists working in the informal industry name or brand their items. This may be detrimental for some reasons, including

- Temporary items are very short-lived, making harvest and use rather limited. This is particularly risky for seasonal plants;
- The change in the strength of the boiled extracts may also be dangerous;
- Low-quality packaging can affect safety and market value;
- Lack of marking and labeling also reduces the value of the product.

9.6 EMERGING ISSUES

The global trade of medicinal plants is increasing at an alarming rate, which poses a major threat to the sustainability of natural resources and ultimately to biodiversity. In addition to traditional medicine, the use of medicinal plants has also steadily increased and diversified in the past few decades, not only in cooking and popular modern western medicine but also in different fields unrelated to medicine, such as the beauty and dye industries (Saha & Sundriyal, 2012). In Europe, for example, there are approximately 2,000 plants that are commercially traded as homeopathic or aromatic.

In recent years, natural medicines have witnessed a revival and expansion in trade volume. Every year, a large number of plant materials obtained from the wilderness enter the global market, and the exchange volume is also increasing. These are usually sent to industrialized countries, which rarely use their own characteristic therapeutic plant assets and gradually restrict the use of some plants. Imports to Europe and North America, including other industrialized countries like Japan come from Asian countries like China, Pakistan Bhutan, Bangladesh, Nepal, Sri Lanka, Hong Kong, India, and African countries including South Africa and Namibia.

Along these lines of thinking, it is obvious that the providing country has made it easier to exchange and consume its medicinal products instead of unfamiliar trade, while the developed countries prohibit the exchange or limited trade of its own species, but may purchases similar plant species from another developing country. Sometimes, shortcomings may lead to a deterioration in the suitability of the plant species or its natural base.

One of the reasons behind the growing threat to restorative plants is to break the standard laws that normally control the use of natural assets. These laws are often effortlessly undermined by current socioeconomic factors. The UK Wildlife and Countryside Act (1981) restricts the uprooting of any wild plant species, except through landowners and other approved individuals. Poland recorded medicinal plant species that were not allowed to be collected without authorization, and Italy passed a law in 1931, which regulated the classification of medicinal and aromatic plants.

On the contrary, in most non-industrial countries, failure to comply with strong guidelines, coupled with poor trade policiey, makes the protection of endangered plant species more difficult. For example, the aromatic stems of Kutki (*Picrorhiza kurrooa*) are highly valuable in Asia, where the base oil is used as a basic fixative for medicines and cosmetics (Schumacher, 2011). Due to trade pressure, the populations of these two plant species have decreased. In non-industrial countries, there is not much drilling for species specification. Usually, a single species is not recorded in the exchange record, the record only contains a collection of families or species. The evidence identifiable at the species level in this way is extremely unreliable, which applies to trade as well.

One of the challenges in regulating global trade is the insufficient labeling of herbal plants for customized purposes. For example, medicinal plants are imported in Bangladesh under the label of "spice". In addition, many factories enter the trade with a code number, which only specifies a product group. These product categories may contain many different plant species; therefore, it is almost impossible to conclude the plant species being traded. Habitat destruction is another major threat to all biological species including medicinal plants. In almost all parts of the world, humans have a major impact on the habitats of florae and fauna-in high mountains and coastal areas, deserts and rain forests the threat of habitat loss, degradation and medicinal plant overharvesting are ever-present.

The destruction of habitat can affect herbal plants in various ways. The most direct impact is the destruction of the ecosystem, which causes the immediate extinction of species in a certain area. The obvious example is the rising cases of "slashing and burning" logging activities in Africa, South America, and Southeast Asia. In addition, once the soil in these areas is barren of natural vegetation, it is susceptible to erosion, especially when not allowed to regrow back within the forest (Shajib et al., 2013). Therefore, the vegetation that originally existed in a given area disappears irreversibly, and the medicinal plants also disappear. In many developing countries, mangroves used for aquaculture, farming or other purposes have been destroyed, resulting in the loss of marshlands and coastal plant varieties. Undoubtedly, the world's climate has been changing at a faster rate for the past 100 years, which is mainly attributed to human activity. With climate change, the change of habitat system is the accompanying effect, which ultimately affects plant species.

9.7 CERTIFICATION LABEL FOR MEDICINAL AND AROMATIC PLANTS

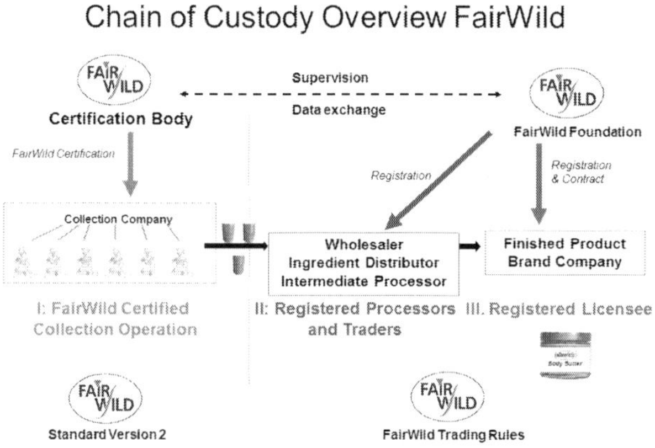

Figure 9.3: FairWild regulates the certification of medicinal plant products.

Source: *https://www.fairwild.org/fairwild-certification*

According to the European Council, organic production can be defined as an independent sustainable management methodology for agriculture and forestry, including the cultivation of plants and animals, production of raw materials, and processing of natural fibers and primary products. It also includes all ecological, economic and socially reasonable production techniques, including the adequate use of fertility of the soil and available water, natural plant characteristics, and the improvement of plant yield and resistance through applying natural forces and laws.

Organic certification organizations can ensure that the collection strategy is feasible and will not harm the biological system and normal production of wildlife. Other non-organic activities, such as WHO's "Good Agricultural and Collection Practices" (GACP) focus on quality control and safety. In addition, the FairWild Standard (FWS) relies on the Global Standards for Sustainable Wild Classification of Medicinal and Aromatic Plants (ISSC Guidelines), which covers competent breed drills, legal and ethical necessities, and competent administrative management and strategic policies. Wild plants are found in many of the things people always use: medicines, cosmetics, wellness products, food, drinks and other household items. For some regions, wild plant collection (WC) forms a major part of their financial activities. At that time, the WC movement is facing some natural and social difficulties due to overharvesting. While various organizations are pursuing this movement, there is a need for proper regulation in the sector.

Fair Wild Standard FWS began in 2006 following the ISSC-MAP guidelines, characterized by determining rules and providing equipment to collectors, manufacturers and decision-makers to manage natural resources following GACP rules. Given modern exchange standards and International Labor Organization (ILO) regulations, as well as mutual implementation with other natural or environmental regulations, the guidelines for medicinal plant use were revised. FairWild's mission is to make natural resource management, and provide strategic guidelines for economically, ecologically and socially viable supply and distribution of the products (Srivastava et al., 2017). FairWild urges all medicinal plant traders to implement the FWS protocol. The FairWild certification is available to manufacturers (to make the life of the target species longer), and to assist brands to use the FairWild trademark to promote their products. Some guiding resources include: FWS overview reports; FWS standards and directions for plant use.

To conclude, medicinal and aromatic plant trading must maintain biodiversity by protecting wild plant species, including beneficial animals that co-exist with these plants in the wild. The role and significance of organic production rests in the feasible management of natural resources, limits the use of chemicals for preservation and fertilizers while establishing harmony with nature. Responsible harvesting of wild medicinal plants encourages biodiversity, promotes the development of communities, ecotourism, and subsequently links agriculture and tourism within a given region (Tao et al., 2009). Future improvements in the medicinal and aromatic plant fields must include: promoting the certification cycle, confirmation of organic production/processing, and checking all applicable permits for the medicinal and aromatic plants.

Chapter 10

THREATS AND CONSERVATION OF MEDICINAL AND AROMATIC PLANTS

CONTENTS

Figure 10.1: Climate change is one of the major threats to aromatic plants.

Source: *https://www.google.com/url?sa=i&url=https%3A%2F%2Fwww.researchgate.net%2Fpublication%2F305749678_Conservation_and_sustainable_use_of_medicinal_plants_Problems_progress_and_prospects&ps*

Development of therapeutic and sweet-smelling plants (MAP) today isn't just a promising other option and contradiction to wild assortment, empowering protection of normal hereditary fluctuation and endurance of uncommon, endemic, defenseless and imperiled species, yet additionally addresses a ground-breaking economy branch giving the fashionable quality crude material for drug, corrective and the food business. Taming and development of the greater part of therapeutic plants, generally considered as a minor yield, face with numerous difficulties on little, medium and enormous scope creation, relating both development advances and market and costs changes. Developed MAP material is progressively liked by the homegrown industry, since it is simpler to foresee plant yield, quality and medication creation, particularly when contrasted and wild gathered crude materials. If there should arise an occurrence of developed MAP material, the chance of plant misidentification and corruption is avoided (Srivastava & Sarkar, 2017). The productivity of development of restorative plants rival benefit attainable for standard field crops for which as of now exist particular hardware and a standard system for utilization of manures and agrochemicals to control weeds, bugs and illnesses. For fruitful huge scope development of MAP, the top-notch crude material ought to be created utilizing low information development techniques to be serious at the global market and with plants gathered from nature (Sayer et al., 2012).

The most well-known issues with which the makers of restorative plants experienced are the market, wealth and openness of wild populaces, agro-natural conditions, work accessibility and expenses, interests in apparatus, post-gather handling, and productivity of creation. Predominant genotypes are vital for the beneficial creation of the excellent therapeutic plants' column material. Out of totally developed therapeutic plant species, just a little rate is plainly hereditarily characterized and addressed on the seed market in term of assortment (Sahoo et al., 2019). Also to different yields, customary reproducing techniques, just as biotechnological strategies and determination helped by atomic markers are applied being developed of new assortments and cultivars of MAP, focusing on the progress of their attractive attributes. This alludes to expanded medication yield and the substance of required auxiliary metabolites. Planning of qualities and specific DNA arrangements engaged with the biosynthesis of specific metabolite classes is by all accounts a future test in MAP rearing projects.

Therapeutic and sweet-smelling plants address a reliable piece of the common biodiversity gift of numerous nations around the globe. These species give a significant commitment to wellbeing, neighborhood economies, social respectability and eventually the prosperity of individuals, especially the provincial poor and delicate social gathering, e.g., the old, kids and ladies. Their job has been progressively recognized in the course of the most recent decade which has achieved another demeanor towards these and different species, differently minimized and ignored by exploration and preservation endeavors. Simultaneously, dangers to the hereditary variety and species endurance have additionally expanded because of impractical abuse and loss of territories. Progress in pharmacognosy has not been joined by equivalent headways in information on the dissemination, hereditary variety, biology and preservation of these species.

Need setting for species determination and comprehension of asset the board needs for these species is compelled by restricted limits and consideration committed to fundamental examination (Shackleton et al., 2007). Latest things in helping to ecological inviting farming and more noteworthy cooperation of neighborhood individuals in the practical preservation and utilization of normal assets are driving towards inventive ways to deal with improve the utilization of restorative and fragrant plants, more participatory in nature and consequently more centered around nearby necessities. The expanding number of associations engaged with restorative and fragrant plants, the admittance to new apparatuses for biodiversity prospecting, portrayal and information investigations alongside the difference in conventional preservation frameworks towards more use-arranged activities ought to be viewed as promising circumstances for updating research objectives and organization, make more prominent cooperative energies at the public level and a more favorable strategy climate. The eventual fate of therapeutic and sweet-smelling plants lays on the present capacity to determine the contentions among preservation and use, and the move towards more asset-based horticulture progressively tested by the globalization of economies.

10.1 DIFFICULTIES ENCOMPASSING THERAPEUTIC PLANTS AND AROMATIC PLANTS

Worries about loss of natural variety and the accessibility of assets. There is no dependable gauge for the quantity of restorative plants that are internationally undermined, differently determined as 4160 or 10 000. There would appear little uncertainty from hypothetical contemplations that numerous therapeutic plant species that have been recorded as compromised, and undoubtedly others that have not, should be experiencing hereditary disintegration now, or will do as such soon. This is on the grounds that populaces of numerous species are in retreat, with remote populaces being annihilated, as the degree and nature of numerous characteristic territories decay. Worries about decreases in nearby information and social endurance. Information on restorative plants, as once installed in a huge number of native societies, is quickly vanishing. Consistently, the aggregate of human information about the sorts, conveyance, environment, techniques for the executives and strategies of extricating the valuable properties of restorative plants is declining quickly – a continuation of a cycle of loss of neighborhood social variety that has been in progress for many years (Singh et al., 2014). There has been an extraordinary development in logical data about therapeutic plants in late many years, however, from various perspectives, this has demonstrated poor pay, because such data is frequently blocked off at field level or is unessential to the issues looked via land directors, or authorities or cultivators of restorative plants.

Concerns identifying with the accessibility and nature of medical care administrations. There is likewise the subject of how best to create customary frameworks to meet present-day challenges. The climate where customary clinical professionals are working today isn't equivalent to in the past. Delicate strategies are expected to maintain a strategic distance from pointless preclusions. Due consideration should be given to conventional guidelines of value, which, in Ayurveda, for instance, traditionally allude to social and tantric use just as helpful characteristics. Concerns identifying with the terms of examination on therapeutic plants. A portion of the worries have emerged in light of information, or doubt, that a few researchers, research foundations or business ventures have taken examples of plants to test for new items, for example, drug, without due consent or on morally inadmissible terms.

10.2 CULTIVATION THREATS FACED BY MEDICINAL AND AROMATIC PLANTS IN INDIA

Pratap Desai summed up the development issues of MAPs in Western Maharashtra, India as follows:

1. Development of MAP: Most of the cultivators don't know about the Agronomy of MAP. Cultivators are new to restorative plants as they know about the development of few conventional harvests like jawar, tobacco, sugarcane. The agronomy of the restorative plants contrasts with conventional plants. In addition, cultivators who know about the agronomy of restorative plants are less in rate.
2. Predominance of bugs and infections: The pervasiveness of bugs in the fields is less, whereas when the yield is reaped and put away, the predominance is more.
3. Environment variety: Most ranchers dealt with the issue of environment variables, as it's a characteristic marvel.
4. Not accessibility of gifted work: Many of ranchers don't get talented works. A restorative plant is another idea for the cultivators and it's hard to get gifted work, as work doesn't know about the restorative plants' agronomy in subtleties.

5. Actual Infrastructure: For the improvement of any association, the physical foundation assumes a significant part. If there should arise an occurrence of MAP the significant framework is the transport office, storeroom, the handling offices and the business sectors for therapeutic and fragrant plants (Tiwari, et al., 2018). Just a couple of cultivators have the office of handling, essential handling like drying and arranging is finished by numerous ranchers and preparing like refining is finished by not many cultivators.
6. Market quality standards of item: ranchers are dealing with the issue to coordinate the quality standards of the item in the Market according to the demand as they don't get preparing of appropriate agronomy of MAP and advertising.
7. Lacking business sector demand data: the cultivators don't get sufficient data about the prerequisite of MAP in the market. As all the plans and data don't reach the farmers.
8. Disorderly market framework: the market framework is profoundly sloppy, there is no fixed example of advertising; the rates and the demand for MAP consistently change.
10. Significant distance to Market place: the commercial centers are far away from the ranches of MAP cultivator, and they can't be able to get the transportation charges that they need to get these products to their target market (Hossain et al., 2017). And other means of transportation are not accessible as ranches are in distant zones which cannot be easily accessed.

10.3 PARTS OF RESTORATIVE PLANT PRESERVATION

The unique meaning of restorative plants in preservation originates from the major social, vocation or financial jobs that they play in numerous individuals' lives. Different arrangements of suggestions have been ordered identifying with the preservation of restorative plants including the requirement for coordinated protection activity, in light of both in situ and *ex situ* methodologies; consideration of local area and sexual orientation points of view in the improvement of approaches and programs; the need for more data on the therapeutic plant exchange; the foundation of frameworks for reviewing and checking the status of restorative plants; the improvement of feasible gathering rehearses; consolation for microenterprise advancement by native and rustic networks; and the assurance of conventional asset and licensed innovation rights.

10.4 NEED FOR USAGE OF THE BOARD PLANS

Restricting the gather to a practical level requires a viable administration framework and sound logical data. The administration framework should incorporate yearly harvest shares, think about occasional or geological limitations and the limitation to collect to specific plant parts or size classes (Legwaila et al., 2011). Likewise, an explanation of the entrance and client rights to the assets giving MAP is important for the fundamental standard data.

10.5 ECO-NAMING AND AFFIRMATION

Given that feasible gathering from the wild is hard to accomplish, affirmation standards can assume a part to guarantee that an item satisfies certain guidelines of maintainability. Accreditation programs identified with characteristic asset use have principally been created for wood and rural items, however, they are by and by being adjusted for wild-gather of non-wood plants. Different plans center around various regions along with the store network: creation, handling, exchange, assembling and promoting. Four classes of accreditation plans have been recognized to be of significance for Guide items: (I) timberland the executives' certificate; (ii) social accreditation; (iii) natural confirmation; and (iv) item quality affirmation.

10.6 EX SITU PRESERVATION, SPREAD, TRAINING AND THE REARING OF YIELD ASSORTMENTS

Plant species can be discovered away from the destinations where they normally happen in a scope of settings, remembering for botanic and different sorts of nurseries, nurseries, seedbanks, tissue culture units, and so on. To act as an illustration of how *ex situ* assortments of MAPs can be associated with preservation and jobs through around situ implies, consider the instance of the Pepperbark tree (*Warburgia salutaris*), the most profoundly valued therapeutic plant in southern Africa (Malik et al., 2013). This species has been gathered to the point of public termination in Zimbabwe, causing challenges in acquiring the medication.

10.7 ENDEAVORS TO NEW ITEM DISCLOSURE

A few phases are engaged with the way toward prospecting the synthetic properties of plants to find drugs or other novel items. This subject is covered legitimately in some new distributions.

The uncommon meaning of restorative plants in protection identifies with the values that they hold for individuals. These qualities concern the commitments that restorative plants can make to medical services, monetary pay, social personality and vocation security. It is accepted that these commitments can here and there be adequately extraordinary that a zero in on restorative plants can possibly accomplish substantially more, regarding preservation and economical turn of events, than only protection of populaces of the therapeutic plants themselves, or practical creation of materials for meds (despite the fact that these, in themselves, are excellent goals). It is valuable to explore the geological connection between the general estimation of therapeutic plants to individuals (if this could be estimated) and the conveyance of plant variety (Mahapatra & Panda, 2012). It is thought that therapeutic plants regularly accomplish their most noteworthy relative qualities in social orders found in spots most extravagant in plant variety. On the off chance that this affiliation is found to hold, at that point projects at numerous focuses of plant variety focused on broad natural preservation could regularly conveniently have an emphasis on therapeutic plants.

10.8 THREATS FACING MEDICINAL AND AROMATIC PLANTS IN INDIA

Five trillion US dollars, that is how much the general global exchange therapeutic plants and their items alone is required to add up to continuously 2050. Evaluations, to the extent therapeutic plants go, are many. As indicated by one by the World Health Organization, 70% to 95% of individuals in many non-industrial nations depend to a great extent on customary medication, generally natural cures, for essential medical services. In 2006, analysts assessed that in excess of 70,000 plant species were being utilized in meds around the world.

Nonetheless, despite their developing monetary significance, therapeutic plants – as an environment administration vital to human wellbeing, business and information – are not mainstream in talk. And with worldwide biodiversity misfortune and disintegration of customary information frameworks, dangers to therapeutic plants and their related information are many. Expanded commercialization of restorative plant assets likewise acquires issues of admittance to this abundance and impartial sharing of advantages. As per the Botanical Survey of India, India is home to more than 8,000 types of therapeutic plants. The nation has a rich history of customary recuperating frameworks, large numbers of which list the utilization of these plants (Mbuvi & Boon, 2009). For example, the most established printed book on Indian restorative plants, Hortus Malabaricus – a 12-volume composition on the therapeutic plants of the Malabar locale along India's west coast – goes back to 1678. Indeed, even in ebb and flow times, research on therapeutic plants and their utilization across India archives how neighborhood networks, particularly native ones, actually use them in their everyday lives.

In the Andaman and Nicobar Islands, four native networks, including the Jarawas, utilize 39 endemic restorative plants for their wellbeing needs and 17 of these are utilized to fix different illnesses, finds a recent report (Mishra et al., 2018). Of the 83 plant species utilized by the semi-peaceful Gujjar people group in Himachal Pradesh, 32 are utilized as medication alone. For sure, numerous networks, particularly native ones, actually depend primarily on customary frameworks of medication for their wellbeing needs, said N Anil Kumar, ranking executive and top of the MS Swaminathan Research Foundation's Community Agrobiodiversity Center at Kerala's Wayanad. Aside from therapeutic plants being indispensable to their occupations, for example, the restorative non-wood backwoods produce that native networks gather and sell professionally, the plants are likewise an essential social asset, he added"These remember non-systematized frameworks, for example, oral accounts for which information is moved by listening in on others' conversations from age to age," said Kumar. Such information frequently differs across networks; once in a while, even across sexes. Kumar describes how a lady from a Kurichiya clan in Wayanad knew about a little, pretty-blossomed spice as a solution for skin rashes for children, while a Paniya man asserted they didn't utilize the spice for any sickness whatsoever.

10.9 A PSEUDOSCIENCE?

In any case, many do scrutinize the logical premise of utilizing direct plant subsidiaries in medication. And not without reason: therapeutic mixtures are famously hard to test utilizing techniques set somewhere near science. A few plant mixes are frequently utilized in a solitary readiness, cleansing and different methods for assessing these mixtures bring about misfortune or change of synthetics. Harmfulness and results, going from queasiness to kidney disappointment, are a chance. Deceptive substances, defilements and substitutes have been accounted for from numerous home-grown prescriptions. In addition, science's brilliant principle of replicability isn't testable, since readiness techniques are regularly intensely protected privileged insights. What can't be denied however is the part of therapeutic plants in medication revelation (Prabha, & Kumar, 2015). As indicated by the WHO, in any event, 25% of all advanced meds are inferred, either straightforwardly or by implication, from restorative plants.

A few medications, including those used to treat malignant growths, are gotten from plants. And science is starting to uncover that some presumed restorative advantages may without a doubt be valid, as it has on account of the arogyapacha spice, *Trichopus zeylanicus*. A few investigations and a nitty-gritty genomic examination later, we could be nearer to understanding why the uncommon "wonder spice" has come to be utilized by the Kani people group of southern Kerala, customarily as a moment energy-releaser.

The difficulties, in any case, such conventional information on networks is quickly disintegrating as there is no exchange of information across ages and more youthful ages are not, at this point intrigued, said Kumar. All the while, human-caused biodiversity misfortunes are influencing restorative plants and related customary information as well. The debasement of woodlands and loss of species make some plant species inaccessible for the utilization of nearby networks. Subsequently, conventional strategies for use including practical techniques for assortment are being lost. Nonetheless, a few endeavors – remembering for situ and ex-situ procedures – are in progress to monitor plant assets and related conventional information (Prabha, & Kumar, 2015). The National Medicinal Plant Board records 72 "therapeutic plants preservation and advancement territories" covering 10,935 hectares that have been set up across 13 states to monitor restorative plants in their characteristic natural surroundings. Under the Department of Biotechnology, four "field quality banks" of undermined plants, including those known for their restorative uses, are additionally being kept up across India.

Extra exploration on such *in vitro* techniques including the creation of auxiliary metabolites would be urgent for the "powerful and practical use" of these plants, remark researchers in an audit of *in vitro* innovation for the protection of therapeutic plants in the Western Ghats. At MSSRF's Wayanad focus, ethno taxonomy – the conventional manners by which networks distinguish and characterize plants – is an investigation center (Pandit & Acharya, 2008). Their six-hectare restorative plant park created with the state government in Odisha's Jeypore capacities as a plant hereditary archive and field site for nine ancestral networks to develop and gather the therapeutic plants they have depended on for quite a long time.

A month-old "cautioning" paper supported by more than 20 researchers overall recommends that environmental change is probably going to influence people groups' admittance to restorative plants. Little taxa, for example, herbaceous plants, particularly in mild zones, would be a portion of the first to be influenced by environmental change, calls attention to K Haridasan, resigned joint head of Foundation for Revitalization of Local Health Traditions, presently The University of Trans-Disciplinary Health Sciences and Technology, who has broadly examined restorative plants and floristics in the western Himalayas and north-east India. A new report alludes to this. In response to poll studies, customary healers in Uttarakhand asserted that a large number of the species they use currently bloom and organic product sooner than previously, which influences their conventional occasions of plant harvests. A few animal groups are currently found at higher heights. Spices that were accessible simply a day's leave their homes are presently more hard to secure, they add.

10.10 CONSERVATION OF MEDICINAL AND AROMATIC PLANTS

Developing a strategy is important in the conservation and Preservation of both medicine and aromatic plants. Conservation of medicinal plants cannot be done by several sectors whether private or public because it requires effort which should include different disciplines and various institutions which will work towards the conservation of medicinal plants and aromatic plants. For the process to work each country should develop a strategy that is national which is suitable for the conservation of these plants and ensure the sustainable use of medicinal and aromatic plants. Preparing a strategy is important because it helps to develop a common goal on what should be done to ensure that these plants are conserved. It will also help in ensuring that the different institutions working together are given specific tasks that they are supposed to ensure that they complete. It will also ensure that those who are participating are very motivated to ensure that they work towards ensuring that these plants are protected. The development of a strategy is also important because it ensures that those who are in charge can track the progress of the conservation methods and results (Saha & Sundriyal, 2012). The guidelines presented acts as checklists of the task to be performed. It is important to note that the roles of different government organizations in different countries are not the same. Preparing strategies is the only way of ensuring that the conservation process is properly undertaken. Starting such a process is quite a hustle but the easiest way of ensuring that the process is successful is by holding workshops that bring people together including experts to assess how severe the situation is, to define the objectives of the institution, to create a list of priorities and coming up with a plan that will be followed by the whole institution. It is also important for group institutions to work with those who use medicinal plants such as herbalists, health workers and the local people. The non-governmental organizations should be present from the beginning.

10.11 STUDYING THE TRADITIONAL KNOWLEDGE ON THE ROLE OF MEDICINAL AND AROMATIC PLANTS

The natural vegetation is disappearing at an alarming rate and many cultures who have lived close with nature depending on nature to provide for their needs are suffering because of the changes that have occurred culturally socially and economically. These people usually have an understanding of the importance of using medicinal plants and aromatic plants but the knowledge is being endangered because of the rapid change that is occurring in such communities. Ethnobotany which is the study of how people use plants to provide for their needs and benefit the world at large is important because it allows the culture to continue and for people to understand and know the role of medicinal and aromatic plants. Ethnobotany is also important because it helped the communities that are going through different changes to adapt to the circumstances that would otherwise prove challenging (Schumacher, 2011). The practice of ethnobotany ensures that the knowledge of the importance of these plans is safeguarded and people benefit from the discoveries that are made from the knowledge they possess. The World Health Organization is encouraging different communities to continue with their practice of using medicinal plants and continue with practices that encourage the use of these plants. This can be achieved by ensuring that each country identifies and supports one or more institutional plans that coordinate and implement surveys that are considered to be ethnobotanical. The other way is to ensure that the ministry of health incorporates traditional remedies that have proven to be of value in the medical field into programs of primary health care.

10.12 IDENTIFYING MEDICINAL PLANTS OUTLINING THE DISTRIBUTION AND ASSESSING THEIR ABUNDANCE

For the medicinal plants to be used effectively it is important to know which species are to be used, their correct names and where they can be found. In many circumstances, the medicinal plants are confused with other plants which have proven to be hazardous especially to the head of a person. Therefore, the health organizations need to identify medicinal plant and gauge whether they are abundant. Every country should ensure that it has well-trained botanists who will be able to identify this plant with enough resources to ensure that the medicinal plants are well identified.

The institutions that practice ethnobotany should ensure that they come up with a catalog that should be used in identifying the medicinal plants. The names of these plants should be in Latin so that they can be easily identified. The geographical distribution should also be included. The catalog should also explain and include the scarcity or abundance of medicinal plants. The habitat should also be included and the description of the appearance of the plant (Shajib et al., 2013). Those plants that are facing extinction should be identified so as to ensure that they are protected by the conservation programs. This is done to ensure that important medicinal plants do not become extinct because if they do, many people who depend on the medicinal plants will suffer. Aromatic plants should also be protected to eliminate the issue of people dealing with the extinction of these plants. The information on the conditional plan should be put on a data basis so as to ensure that the authenticity of the information obtained by other researchers is conclusive and true. This is to ensure that the information obtained is not biased and can be trusted.

10.13 AT EVERY POSSIBLE OPPORTUNITY, TO PLANT AND CULTIVATE MEDICINAL PLANTS AS SOURCES OF MEDICINE TO BE USED BY PEOPLE

The most ideal approach to give the plant material expected for medication is to grow plants that have medicinal value. It is considered to be better than gathering the plant materials from wild resources because it ensures that it doesn't exhaust stocks that are wild, and in numerous cases, the territories of medicinal plants that are don't supply the growing business sector for therapeutic plant items. On account of uncommon, imperiled or over-abused plants, development is the best way to give material without further imperiling the endurance of those species that belong to the medicinal plants. Development additionally has pharmacological preferences over wild-gathering. Plants that are wildly gathered ordinarily change in quality and piece, because of ecological and hereditary contrasts. In development, this variety - and the subsequent vulnerability of the restorative advantage - is tremendously diminished (Shajib et al., 2013). The plants can be filled in territories of comparable environment and soil, they can be flooded to expand yields also, they can be harvested at the perfect time. Development likewise enormously decreases the chance of misidentification and adulteration.

Numerous nations have long conventions of developing therapeutic plants. For instance, in certain nations, housewives customarily grow a scope of spices fundamental for the medical services of their families in pots which are then situated around their homes. However, in others, the therapeutic plants are obtained from the wild by people who understand what they are doing; as the population of human increases, as timberlands decay and as territories of staying wild living spaces become secured in stops and saves, individuals in certain territories cannot, at this point gather the plant materials they need (Malik et al., 2013). The need to build up the custom of growing and coming up with methods of growing the plants is basic. Some combination of guidelines and motivators might be expected to empower development, particularly in pieces of the jungles where numerous therapeutic plants are moderate developing trees.

The Ministries of Agriculture and Health should ensure that they come up with a program for building nurseries where these plants are grown. The World Health Organization program that involves medicinal plants and traditional Medicine Program has plot two fundamental systems for the creation and the use of therapeutic plants: (a) the agro-mechanical creation and utilization of restorative plants of normalized pharmacologically dynamic constituents; and (b) the circulation of seeds, seedlings or potentially saplings to people and networks for development in gardens that are found at home. For instance, plants that have proven to have properties that cure and treat malaria could be conveyed to networks during the season when malaria is mostly transmitted.

10.14 THOSE PEOPLE WHO PLAN THESE PLANTS SHOULD RAISE MORE UNIFORM, HIGHER YIELDING TYPES OF RESTORATIVE PLANTS THAT ARE CONSIDERED TO HAVE MEDICINAL PROPERTIES

Plant reproducing has been utilized for a long period of time in the growing of plants to deliver improved types of plants that are called cultivars, which are adjusted to the growth process. Be that as it may, so far it has simply been applied to not many therapeutic plants. Plants that are reproducing cannot just improve the agronomic attributes of a therapeutic plant by developing medicinal properties, for example, by upgrading the genetic composition of a certain plant which improves its protection from bugs and sicknesses, yet in addition improve the phytochemical attributes of the plant, quite to build up a high and uniform convergence of the dynamic constituents that allow it to have the medicinal properties. Most plants that have medicinal properties are at present developed as unchanged wild plants, thus will in general be truly factor, indeed, even in the growth process. Plant reproducing can decrease this changeability, thus help normalize the characteristics that are considered to be therapeutic. Additionally, plant reproducing can adjust the plants for development in various soils and environments, so that the plants would then be able to be developed under a wide scope of conditions. Powerful plant rearing has to involve a number of things but the most important is the genetic and hereditary variety in the plant as beginning material which can either be obtained from wild stocks or from *ex situ* protection Botanic cultivates specifically have a significant part in keeping up wide genetic stocks of therapeutic plants for plant reproducing. Notwithstanding, the expensive and relentless methods of present-day plant rearing are simply liable to be utilized in the instances of plants developed for high-volume crops required for present-day medication (Mahapatra & Panda, 2012). The venture required is probably going to be excessively high for a large number of the plants utilized in customary frameworks of medication. For this situation, a public program for plant reproducing should think about their necessities.

Development practices ought to limit the utilization of synthetic compounds. The utilization of weed-executioners and pesticides may address a danger to the climate, to those becoming the plants and to those treated with the subsequent medication. Along these lines, the utilization of synthetic compounds (whenever utilized by any means) ought to be diminished to a flat-out least, and where synthetic substances are utilized, essential guidelines ought to be set up and inflexibly followed. Botanic gardens and horticultural institutes should provide effective horticultural training and information. Provision for training and information is vital (Mbuvi & Boon, 2009). Training may be the best approach for those working in medicinal plant nurseries. Information leaflets and other forms of extension may be more suitable to help members of the public grow their basic home pharmacies themselves.

INDEX